# CONGENITAL ARTERIOVENOUS ANEURYSMS OF THE CAROTID AND VERTEBRAL ARTERIAL SYSTEMS

BY

## H. OLIVECRONA
STOCKHOLM

## J. LADENHEIM
STOCKHOLM · NEW YORK

WITH 122 FIGURES

SPRINGER-VERLAG BERLIN HEIDELBERG GMBH

1957

PROFESSOR H. OLIVECRONA, M. D., SERAFIMERLASARETTET, STOCKHOLM
DR. J. LADENHEIM, BOX, WEST NEW YORK / NEW YERSEY (USA)
1200 EAST BROAD STREET, RICHMOND/VIRGINIA

ISBN 978-3-540-02204-6    ISBN 978-3-662-11367-7 (eBook)
DOI 10.1007/978-3-662-11367-7

# Contents

# Introduction

Two decades have elapsed since the publication of the earliest study from this clinic concerning arteriovenous aneurysms. Although numerous communications[13, 36-40] have elaborated on our subsequent experiences with this lesion, it would appear justified and profitable at this time to further review the material and assess the results. Moreover, we shall attempt in this monograph to concentrate primarily upon our own clinical material, the largest series published to date, rather than prepare an exhaustive survey of the literature. Attention is invited to the study of TÖNNIS and LANGE-COSACK[63], with which this series may be collated.

Acknowledgements are herein made to other members of the clinic who have previously reported on this subject, from whose studies we have in several instances reproduced case histories. We express our thanks for the assistance rendered by Professor E. LINDGREN. Drs. H. NORDENSTAM and O. HÖÖK were kind enough to place their services and advice at our disposal. To the Springer Publishing Company goes the gratitude of the authors for the interest, effort and cooperation attending the publication of this monograph.

# Historical Review

Fleeting allusions to the signs of arteriovenous aneurysm are contained in the writings of Antyllus (Second Century), Albucasis (Tenth Century), Vidus Vidius (Seventeenth Century) and Sennertus (Seventeenth Century)[41, 66]. It remained for WILLIAM HUNTER (1757)[41], however, to present the earliest recorded cases in which the abnormal communication between artery and vein was identified. In addition to recognizing the significance of the thrill, HUNTER demonstrated that compression applied in the region of the fistula produced both a diminution in the size of the peripheral vessels and an elimination of the murmur.

Application of the Hunterian concept to the problem of congenital arteriovenous aneurysms was initiated with the classic observations of WRIGHT (1831)[46] and BELL (1832)[46, 66] regarding the vascular composition of the lesion. The postmortem dissections of STANLEY (1853)[46] and WARREN (1837)[46] demonstrated a direct communication between artery and vein. KRAUSS (1862)[66] performed injection studies of the lesion in an amputated limb. The anastomatic factor was identified *in vivo* by WERNHER (1876)[46] who, after excising the center of the lesion, noted a diminution in size and pulsation of both afferent and efferent vessels. The terminology of these malformations —variously reported as tumor circoidius, varix aneurysmaticus, aneurysma per anastomosin, angioma arteriale, angioma plexiforme, varix arterialis, aneurysma cirsoidea, aneurysma serpentina, aneurysm racemosum, etc.—was simplified by BRAMANN (1886)[46] who classified these entities as arteriovenous aneurysms, but so great was the residual confusion that MELENEY (1923)[34], in compiling an index of angiomatous malformations, neglected to include the German literature where the lesions had been listed as Rankenangiom.

The earliest clinical observations on the congenital arteriovenous aneurysm occurring in the brain substance were reported by STEINHEIL (1895)[59], HOFFMAN (1898)[19] and ISENSCHMID (1912)[25]. An excellent clinico-pathological appraisal of the intracerebral lesion was presented in 1928 by CUSHING and BAILEY[7]. Shortly thereafter, DANDY[8] published a study of his surgical experiences with 8 patients, but, unfortunately, could not report a single survival where a complete extirpation had been attempted. In 1936 BERG-STRAND et al.[3] compiled a classification of the angiomata which to date has retained its usefulness for surgical orientation, and in the same monograph a report was presented of 5 congenital arteriovenous aneurysms successfully extirpated in their entirety.

# Classification

## 1. Etiologic Classification

The arteriovenous aneurysm may be defined as an abnormal, direct communication between arterial and venous channels without the interposition of a capillary system. Among the many etiologic classifications suggested, a separation of this entity into congenital and acquired forms appears to best facilitate description (Table 1).

Table 1. *Classification*

| Etiologic | Pathologic |
|---|---|
| 1. Acquired | 1. Cavernous |
| 2. Congenital | 2. Racemose |
|    a) Anomolous arteriovenous |    a) Telangiectasia |
|    b) Angiomatous arteriovenous |    b) STURGE-WEBER |
| |    c) Venous racemose |
| |    d) Arterial racemose |
| |    e) Angiomatous arteriovenous |

**Acquired arteriovenous aneurysms.** These are abnormal direct communications between the arterial and venous systems associated with trauma or disease within, on, or around the walls of the artery. Often the nature of the trauma may be obscure, as, for example, in the carotid-cavernous aneurysms which appear shortly after birth, where neglect of the birth trauma factor has led to the erroneous classification of the lesion as "congenital". There can be little question of the existence of a congenital predisposing factor in many of these acquired lesions, especially in the carotid-cavernous variety. Studies since the time of Charcot[10, 11, 50] have demonstrated an inherent weakness of the carotid artery, situated at the site of the primitive trigeminal artery[42, 61]. Notwithstanding these considerations of predisposition, careful questioning of the patient will almost always uncover evidence of superimposed trauma, often in a seemingly insignificant form (sneezing, partuition, etc.), justifying the classification of these lesions as acquired.

**Congenital arteriovenous aneurysms,** the category which will concern us in this study, is of the anomalous and the angiomatous varieties.

*Anomalous arteriovenous aneurysms* are exceedingly rare. As the name implies, these arteriovenous aneurysms are produced by the anomalous passage of a small artery or vein into the "wrong" vascular channel. REINHOFF[49] has recorded a direct arteriovenous anastomosis of anomalous external jugular branches with the external carotid artery. Similarly, a case reported by DANDY[9, 10] describes the anomalous course of a branch from the middle cerebral artery into a large venous trunk, ultimately reaching the sylvian vein. No such lesions were encountered in this series and, accordingly, further mention will not be made of this entity throughout the remainder of this study.

*Angiomatous arteriovenous aneurysms.* These are congenital malformations which in previous years have masqueraded under a variety of titles. Unlike the anomolous arteriovenous aneurysm, where the communication between arterial and venous channels is effected by a single or relatively few vessels of normal maturity, the angiomatous arteriovenous aneurysm communicates via many vessels, all demonstrating histologic immaturity or malformation.

## 2. Pathologic Classification

The angiomatous arteriovenous aneurysm, together with several other angiomatous entities, comprise the racemose group of angiomata, characterized by the presence of parenchyma between the vessels of the malformation. In contrast, parenchyma is absent in the cavernous angioma, and the vessels lie in juxtaposition to one another.

**Racemose angioma.** Among the racemose angiomata, in addition to the angiomatous arteriovenous aneurysm, several other forms must be briefly discussed, for their physiologic similarities often obscure an accurate identification at the time of surgery.

*Telangiectasia and* STURGE-WEBER'S *Disease* have little significance for this discussion. These malformations of capillary or precapillary blood vessels are well defined clinical and pathological entities that rarely, if ever, enter into the differential diagnosis.

In the *arterial racemose angiomata* (Fig. 1) both feeding and draining vessels should *a priori* have an arterial composition. While such a lesion is conceivable, a convincing example has yet to be recorded. Such lesions as have in the past found their way into the literature under this name were most probably angiomatous arteriovenous aneurysms.

*Venous racemose angiomata* (Fig. 1) are rare, but a few authentic cases have been described in the brain and spinal cord. In these angiomata both feeding and draining vessels are of venous construction. CUSHING and BAILEY[7] have directed attention to the close relationship of this lesion to the arteriovenous aneurysms, while BERGSTRAND et al.[3] described them as histologically indistinguishable. Such differences as exist are of physiologic nature, predicated upon the extent of the arterial blood supply. Thus, if the amount of arterial blood is sufficient to impart an arterial color to the blood passing through the malfor-

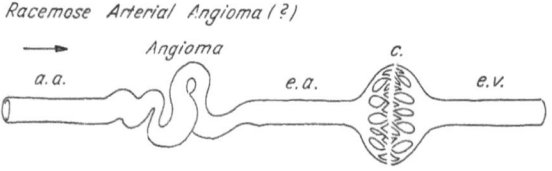

Racemose Arterial Angioma (?)

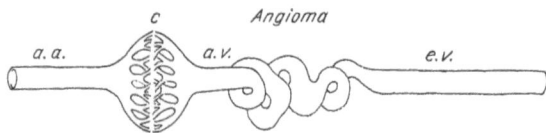

Racemose Venous Angioma

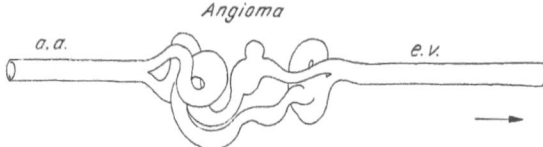

Angiomatous Arteriovenous Aneurysm

Fig. 1. The racemose angiomata. (Drawn by Dr. J. MORA-RUBIO after LIMA.) a.a. afferent artery, e.a. efferent artery, a.v. afferent vein, e.v. efferent vein

mation, the lesion should be classified as an arteriovenous aneurysm, while if the arterial supply is sparse, the venous hue will predominate and the lesion can then properly be classified as a venous racemose angioma. Since this distinction can be made only *in vivo*, the suspicion may be justifiably entertained that most of the cases described in the literature as venous racemose angiomata, where postmortem examination alone was conducted, were in reality arteriovenous aneurysms.

The *angiomatous arteriovenous aneurysm* (Fig. 1) is by far the most common, and clinically the most important, form of racemose malformation. The embryonic agenesis of the capillary system belonging to an artery or a group of arteries results in the discharge of arterial blood directly into the venous system through a tangle of abnormal blood vessels of varying caliber which join the two channels. The veins draining this mass of blood vessels are dilated and pulsate with each systole, while the arteries which feed the abnormal communication are engorged and dilated. Evidence of either arterial or venous characteristics may at time be discerned in the small vessels comprising the lesion, but more often they are undifferentiated and pathologic in structure.

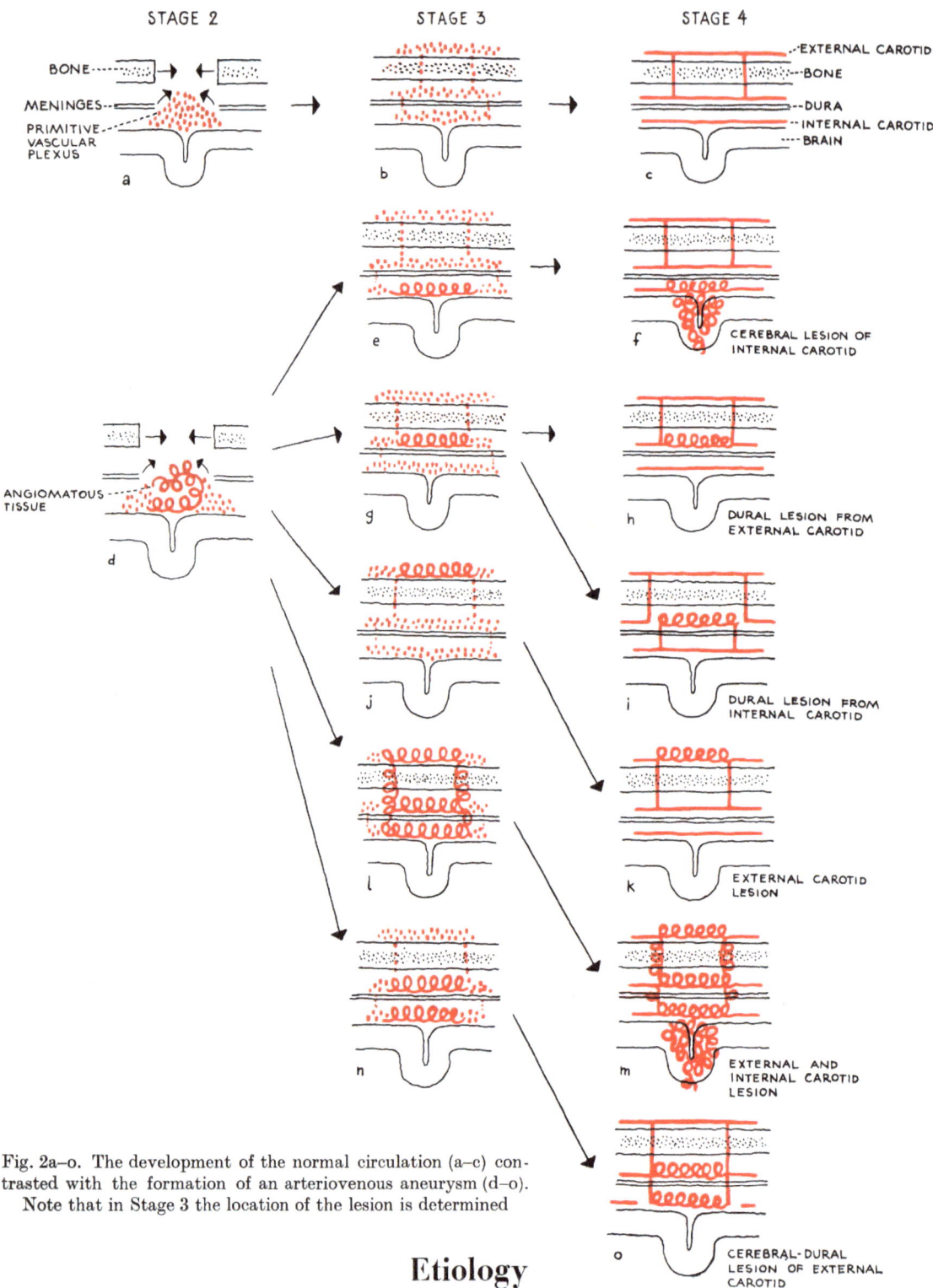

Fig. 2a–o. The development of the normal circulation (a–c) contrasted with the formation of an arteriovenous aneurysm (d–o).
Note that in Stage 3 the location of the lesion is determined

# Etiology

Since artery and vein within the brain substance do not course in juxtaposition, trauma cannot be accepted as a causative factor in the intracerebral arteriovenous aneurysm. This early explanation has now been permanently laid to rest, with the publication of a pedigree of lesions occurring in siblings[29].

The etiology of these angiomata, long the subject of controversy, has in recent years been almost universally attributed to developmental malformations of the cerebral vascular system. So suggestive is the pattern of embryological development, that although little is known of the formation of the lesion, it is difficult to refrain from speculation.

The period of early development of the cerebral circulation has been traditionally divided into five chronological stages[1, 42, 61].

*1. Primitive vascular plexification.* Angioblasts differentiate to form solid masses of cells. The clumps, initially aligned into band- or cord-like structures, gradually assume a plexiform configuration.

*2. Vessel formation.* Vacuolization occurs within these solid structures, resulting in the formation of a lumen. Around the cavity, the peripheral cells differentiate to form endothelial cells which, in turn, further produce endothelium and other angioblasts. The vascular plexus, appearing first over the most cephalad portion of the brain, soon extends over the remaining brain structures, evolving into afferent, efferent and capillary components. The capillary bed retains an adherency to the surface of the brain while the superficial part of the plexus forms channels of large and intermediate caliber which maintain a communication with the capillary bed.

*3. Stratification.* With the development of the meninges and cranium, the vascular system, commencing at the base of the brain and slowly spreading to the vertex, becomes stratified into superficial, dural and pial vessels.

*4. Rearrangement.* The primitive vascular channels become obliterated as newer branches form to replace the early pattern. It is during this stage that the trigeminal and hyoid-stapedial arteries undergo their involved changes.

*5. Histological development.* With further morphological development of brain structure, histological differentiation procedes within the walls of the artery, vein and capillary bed.

Since the predominating characteristic of the congenital angiomatous arteriovenous aneurysm is focal capillary agenesis, it would not seem unreasonable to assign the development of the lesion to the period characterized by the formation of the capillary bed, i.e. *Stage 2.* Accordingly, the mechanics of *Stage 3* will determine whether the arteriovenous aneurysm will be situated in the scalp, brain or dura; and by *Stage 4*, the vessels feeding the malformation will be delineated. This hypothesis is presented in schematic form in Fig. 2. Inasmuch as the stratification of the meninges is most likely to be retarded over the lateral aspect of the brain[65], an explanation of the frequent occurrence of communications between external carotid and middle cerebral arteries can be postulated.

# Physiopathology

Despite the multiplicity of causes and the variety of types, the arteriovenous aneurysms have a common physiological denominator. Accordingly, the discussion will not be too greatly burdened by an abstraction of the changes resulting from an arteriovenous aneurysm of moderate size established somewhere in a large channel of the peripheral vascular system (Fig. 3 B).

Inasmuch as the peripheral resistance from fistula to heart via vein ($v'$) is considerably less than either the resistance of fistula to capillary bed via vein ($d'$) or the resistance of the normal pathway ($b + c$) existing prior to the establishment of the fistula, arterial blood will be routed to the site of fistula and preferentially shunted through the communication to the venous system leading to the heart. This abnormal diversion of flow having been established, certain compensatory changes occur in the vessel walls in order that they may better withstand the burden of their newly acquired tasks. Deprived of the strain upon its walls as a result of the reduction of peripheral pressure, the proximal feeding artery ($a'$) undergoes degenerative changes. The walls become thinned, the muscle layer atrophies, the elastic tissue decreases in quantity and, in certain instances, the artery becomes tortuous. This has been described as a "venification" for, indeed, the function of the vessels leading to and from the fistulous area approach a physiologic similarity. On the other hand, in response to increased pressure and work demands, the draining vein ($v'$) "arterializes", with the hypertrophy of its muscular layer. As the pressure gradients between the distal artery segment ($b'$) and its anastomosing vessels become marked, collateral vessels develop. The heart—now saddled with the twofold burden of *1*, pumping blood through smaller channels having a greater resistance in an effort to supply the peripheral organs and *2*, pumping "unconsumed" blood through a parasitic channel—hypertrophies to increase the stroke volume output. Pulse rate accelerates, the circulation time diminishes and blood volume is increased. While these changes are in progress, the draining vein to the heart ($v'$) and, to a lesser degree, the veins leading to the fistula ($d'$), dilate to accomodate the new volume of blood, but this development is attended by increases in the peripheral resistance within the aforementioned components. If the fistula is assumed to have the property of perfect elasticity, the fistula will also dilate. The proximal artery too, for reasons not clearly understood, will undergo dilatation. Eventually peripheral resistance in the abnormal channels increases to the extent that it equals that of the normal channels ($c + v = f + d' + v'$). At this point, a state of equilibrium is reached and cardiac hypertrophy will not progress further. If, however, this state is not attained before the hypertrophy resources of the heart have been exhausted, cardiac dilatation, heart failure and death will intervene prior to establishment of an equilibrium in resistances[16, 20, 21, 25, 27, 45, 69, 70].

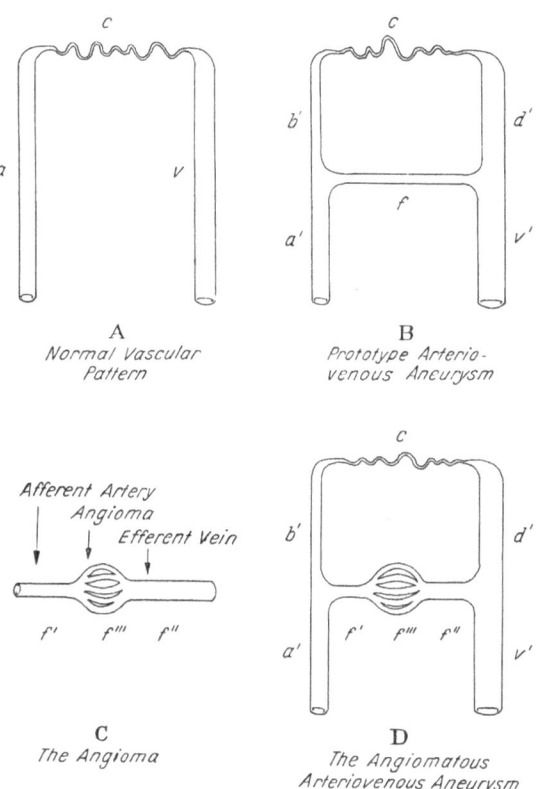

Fig. 3A–D. The components of a normal capillary system, an arteriovenous aneurysm and a congenital arteriovenous aneurysm (see text). *a* normal artery; *c* normal capillary bed; *v* normal vein; *a'* physiological feeding artery; *b'* distal artery; *d'* distal vein; *v'* proximal vein; *f* fistula; *f'* afferent artery; *f''* efferent vein; *f'''* angioma. Note that the true feeding artery (*a'*) is not the afferent vessel (*f'*), but the artery which supplies the afferent vessel

Because of certain unique characteristics of the congenital arteriovenous aneurysm and the peculiarities of the cerebral circulation, the sequence and degree of change, somewhat arbitrarily stated above, undergo modification when the schema is applied to the cerebral blood flow. We shall attempt to analyze these factors as they pertain to the congenital arteriovenous aneurysm of the carotid and vertebral systems.

**1. Comparative structure.** The functioning arteriovenous aneurysm, as mentioned above, is composed of proximal arterial ($a'$), distal arterial ($b'$), proximal venous ($v'$), distal venous ($d'$) components and the point of abnormal connection or fistula ($f$). A certain confusion has crept into the literature when the term "fistula" is used synonymously with arteriovenous aneurysm and for this reason it is particularly desirable to limit its use exclusively to the communicating channel.

At first inspection (Fig. 3C) it would appear that since the congenital arteriovenous aneurysm is a direct communication in the absence of an identifiable capillary network, both the distal arterial and distal venous components of the congenital shunt should be lacking. Further reflection (Fig. 3 D) suggests that distal venous and arterial factors are represented in the normal branches of artery and vein which originate from the trunks which supply the afferent artery and efferent vein to the angioma. The afferent artery ($f'$) of the fistula is not, therefore, the physiologic feeding artery ($a'$) of the congenital arteriovenous aneurysm, but, together with the efferent vein ($f''$) and the malformed vessels ($f'''$), constitutes the fistula ($f = f' + f'' + f'''$). This concept purports to explain why the area of compromised circulation, in the presence of a congenital arteriovenous aneurysm, is not limited to the region surrounding the angioma supplied by the afferent artery ($f'$), but involves a larger area supplied by the other branches of the trunk from which the afferent artery to the angioma orginates.

**2. Size and number of fistulae.** The greater the caliber of the communicating channel, the greater the total volume of blood that can be diverted directly into the vein. Although the congenital arteriovenous aneurysm is composed of a number of communicating vessels of varying patency ($f'''$), the true dynamic caliber of the fistula is the diameter of the afferent artery ($f'$). The maximum amount of blood transmissible is a function of the caliber of the trunk which supplies the afferent artery, i.e. the physiologic feeding artery ($a'$).

**3. The feeding artery ($a'$).** The paucity of branches of the carotid artery between its origin from the arch of the aorta to the Circle of Willis is a phylogenetic insurance that the brain of man will receive ample nourishment. Thus, the relatively small size of the external carotid artery—which has diminished *pari passu* with the atrophy of scalp and face musculature and the loss of anastomotic connections (rete mirabile) to the intracranial division of the internal carotid artery—has resulted in the transmission to the human brain of arterial blood under great pressure. Indeed, blood pressure in the cortical arteries can attain a value of 80% of the common carotid pressure, thereafter falling abruptly to a level of 20 mm. in the region of the capillary bed[15, 60]. Since the physiologic feeding artery ($a'$) is often a principal branch of the Circle of Willis, it becomes apparent that high pressures are involved in the congenital arteriovenous aneurysm, despite its peripheral location.

At the beginning of this section it was noted that the feeding vessel may, for reasons unknown, dilate in the course of development of the abnormal communication. Several theories have been offered to explain the fusiform dilatations seen with the traumatic arteriovenous aneurysms (vasa vasorum alterations, interference with the normal pulsations necessary for the integrity of the vessel wall, etc.), but the matter remains to date somewhat obscure, even after a controversy of three decades.

This phenomenon of dilatation of $a'$ is of interest for, in association with arteriovenous aneurysms of the brain, saccular aneurysms of the circle of WILLIS or its immediate branches have been reported (LEY GARCIA and OLIVERES: Annual Meeting of the Society of British Neurological Surgeons, 1956). Although this occurrence has not been seen in our own material, we wish to invite attention to a possibly analogous relationship between the two forms of dilatation associated with the arteriovenous aneurysm.

**4. Differences in peripheral resistance** between the normal circuit and the parasitic circuit. The closer the aneurysm to the heart, the less the effort required to propel the blood through the abnormal channel, for the negative tension *(vis a fronte)*, created by the sucking action of the chest, increases in the direction of the heart at a rate of about 1 mm./35 mm. of distance[15]. In this respect the cerebral malformation, because of its peripheral location is confronted by greater peripheral resistance in $v'$ than the acquired variety of the cervical region, but even in so distal an organ as the brain the decrease of peripheral resistance is formidable, measuring in some instance $1/3$ of normal[55, 56].

**5. Proximal branches of the feeding artery.** Since mention of the effect of the fistula on the branches of the feeding artery proximal to the fistula has not thus far been made, it may be well at this time to review this matter in some detail.

The tortuosity, thrombosis, intimal proliferation and focal occlusion of the pathologic vessels of the angioma are among the sources of peripheral resistance at $f$.

Where the proximal branches ($k$) are situated close to the heart (Fig. 4), the peripheral resistance, for reasons mentioned above, will be negligible and certainly less than at $f$, so that the supply of arterial blood to $k$ will be relatively unchanged. When $k$ is situated closer to the site of fistula, the peripheral resistance increases, and at a certain point, it will exceed the resistance at $f$. When this occurs, the resistance in the arterial branches proximal to the fistula will have virtually the same relation to $f$ as in the distal artery ($b'$), i.e., the peripheral resistance of both $b'$ and $k$ will be greater than the resistance at $f$, although, to be sure, the resistance of the $k$ circuit will always be less than that of $b'$.

To state the matter more concretely, if the malformation is situated on a branch of the middle cerebral artery, a significant alteration of circulation should be demonstrable in not only the area of distribution of the middle cerebral, but the anterior and posterior cerebral arteries as well. This hypothesis is supported by clinical experience. "Sucking action of the arteriovenous aneurysm" is a clinical description of the preferential diversion of blood flow through the shunt at the expense of the other vessels in the proximity of the communication.

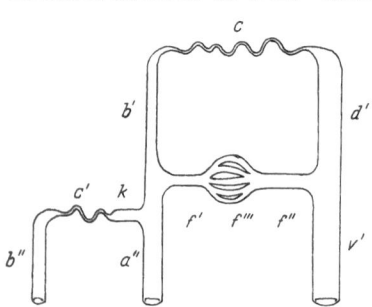

Fig. 4. The normal vascular branches proximal to the fistula (see text). $k$ an arterial branch proximal to the fistula; $c'$ capillary system proximal to the fistula; $b''$ vein proximal to the fistula

**6. Duration.** The longer the duration of the aneurysm, the greater will be the evidence of compensatory changes on the part of the body to overcome the disadvantages resulting from blood flow alteration. Nevertheless, cardiac function studies performed at the Serafimer Hospital in 9 patients with congenital arteriovenous malformations revealed no significant changes which could be related to the intracranial lesion[22]. That congenital arteriovenous aneurysms of the brain cannot be implicated in marked disturbances of cardiac function does not invalidate our premise, for, because of the fragility of the blood vessels of the malformation, a "blow-out" of the fistula (intracranial hemorrhage) will precede marked cardiovascular alterations. This is a fundamental difference between the congenital and the acquired arteriovenous aneurysm.

Many patients, to be sure, have survived for decades *after* intracerebral bleeding without demonstrating evidence of cardiovascular disturbance. This observation may be explained by postulating an alteration of the lesion following intracerebral bleeding. In this series complete destruction of the malformation after hemorrhage was encountered in one instance and partial destruction has been often observed at surgery. Therefore, in proportion to the destruction of the lesion, the probabilities for cardiovascular derangement are diminished following hemorrhage, while the likelihood of a subsequent episode of hemorrhage, on the other hand, will depend on the nature of the alteration of the lesion. As long as functioning angiomatous tissue remains, however, it is probable that another vascular accident will precede cardiovascular decompensation.

**7. Reaction around the fistulous area.** As indicated in the general outline, a fistulous communication would—assuming perfect elasticity of both it and the venous channels—continue to enlarge indefinitely until a state of equilibrium or cardiac dilatation was attained, were it not for the reactivity of the surrounding tissues and their ability to restrain vessel dilatation. Communications of the external carotid artery occurring in the scalp are contained in a capsule-like enclosure formed by reaction of the surrounding tissues (Fig. 5A), while the intracranial arteriovenous aneurysm is circumscribed by the gliosis in the adjacent brain structure (Fig. 5B) and by pathologic changes within the walls of the vessels (hyalinization, calcification, etc.) which diminish their elasticity.

**8. Distensibility of the peripheral tissues.** The greater the distensibility of the tissues, the greater the potential diversion of arterial blood. This factor is of especial importance for the brain, for, allowing for a limited tampon effect of the cerebral spinal fluid, the cranial cavity, in comparison with other organs, is virtually inexpandible. Intracerebral resistance to dilatation of $d'$ and $v'$ is therefore maximal, far exceeding the mild restraint offered by galea or facial tissues.

**9. Collateral anastomosis.** The more extensive the arterial network peripheral to the site of abnormal communication, the less the damage to the distal tissues. Hypothetically, with the development of a rich anastomosis peripheral to the fistula and with adequate cardiac compensation an undiminished blood supply might be provided to the peripheral tissues. Although tissue nutriment is, in fact, relatively unimpared in lesions of the external carotid artery, such does not obtain with congenital arteriovenous aneurysms of the brain.

Collateral vessels do not appear from nowhere, of course, but must be present prior to the creation of the fistula, for the maximal capacity for collateral development is limited by the preexisting anatomic pattern. The small arteries of the brain are, in effect, *physiologic* end-arteries, despite the fact that structural connections may, or may not, exist at the precapillary level or in the so-called "capillary continuum"[73]. The net effect of this paucity of effective collateral possibilities peripheral to the site of communication is pemonstrated by angiogram where a lack of filling of a large segment of the arterial structure within the brain is often demonstrable in the presence of an angiomatous malformation (Fig. 49, 50).

Diminished circulation implies diminished nutrition to the tissues. Because of the high metabolic rate of the brain, circulatory and nutritive aberations have a particularly harmful effect on cerebral tissue. The brain, comprising 2 % of the total body weight, receives 17 % of the cardiac output and consumes 22 % of the total oxygen and 70 % of the liver glycogen[27, 52]. 1. *Arteriovenous oxygen studies.* If the arterial and venous oxygen values are measured, the differences in the presence of an arteriovenous malformation are found in some instances to be $1/3$ of normal. The profound loss of oxygen implied by this proportion, moreover, is only a single expression of the diversion of vital nutritive substances into the parasitic circulation.

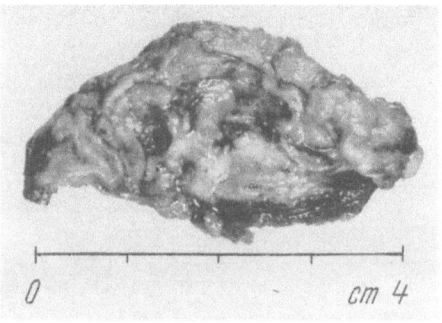

Fig. 5A. Operative specimen. J. B. E., 10 year old boy, 901/53. A lesion of the superficial temporal artery. Note the pseudocapsulation of the lesion

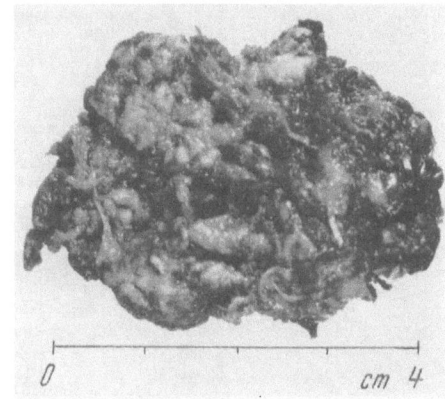

Fig. 5B. Operative specimen. A lesion of the left parietal region in a 29 year old woman (B. R. J. 877/53). The pseudocapsulation of the lesion was formed by the gliotic reaction of the brain parenchyma

2. *Radioactive iodine studies.* The rapidity of blood flow through the abnormal communication can be measured by radioactive iodine studies[74]. Not only does the isotope appear pathognomonically promptly in the venous circulation, but the relative height of the venous peak appears to be greater than that observed in the normal circulatory pattern. The rapid and premature appearance of all nutritive substances in the venous system at the expense of cerebral consumption may be presumed to follow patterns somewhat related to the above.

**10. Nature of the fistula.** If the fistulous connection lacks the histologic integrity to withstand an increase of pressure, it will, in proportion to its weakness, constitute a *locus minoris resistantiae* to increases in cardiac output and other stress. Not only does the congenital arteriovenous aneurysm have embryonic deficiencies in the muscular layer which limit the capacity for hypertrophy, but certain pathologic changes may subsequently occur in the vessels which further accentuate the danger of rupture. In short, the pathologic vessels ($f'''$) cannot keep pace with the ability of the other components of the fistula ($f' + f''$) to dilate.

# Pathology

*Gross inspection* of the malformation (Fig. 6) summons to mind the expression "hemorrhoids of the brain". The lesion is composed of a snake-nest of tortuous, fragile, arterialized vessels which originate from the pial vascular system. Regardless of the

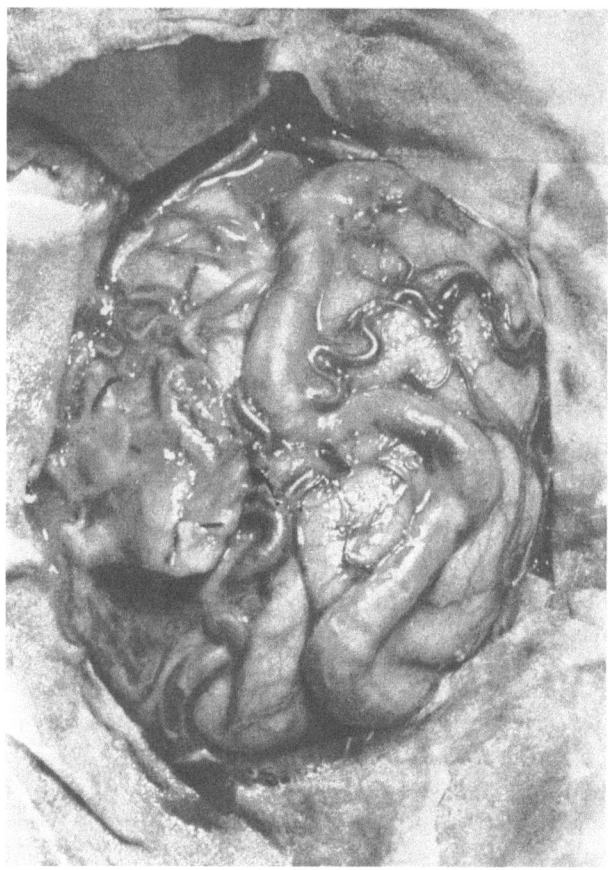

Fig. 6. Appearance of the arteriovenous aneurysm at surgery. E. H. A., 45 year old woman, 776/49

abundance of abnormal vessels presenting on the exterior, it is axiomatic that a larger mass will be found beneath the brain surface, projecting to the interior in the manner of an inverted cone, with the vessels of the apex directed to, and frequently communicating with, the vessels of the choroid plexus (Fig. 7, 28). The dilated afferent system communicating with the angiomatous cone may be difficult to identify on the brain surface but the swollen and somewhat fragile veins of the outlet are always in evident abundance.

The *microscopic* appearance of the congenital arteriovenous aneurysm is bizarre, reflecting the degrees of embryologic maldevelopment (Fig. 8, 9). The aneurysmal bed is composed of a tangle of blood vessels of varying caliber, thickness and structure. Hyalin degeneration, evidence of previous hemorrhage, calcification, and thrombosis can at times be observed. The proliferation of *intima* of the smaller vessels replaces a well-defined elastica interna to the extent that occlusion of the lumen may occur in certain areas, while in other places only scattered strands of elastin are discernible. The *media* is composed of a poorly developed muscle layer especially deficient in the longitudinal

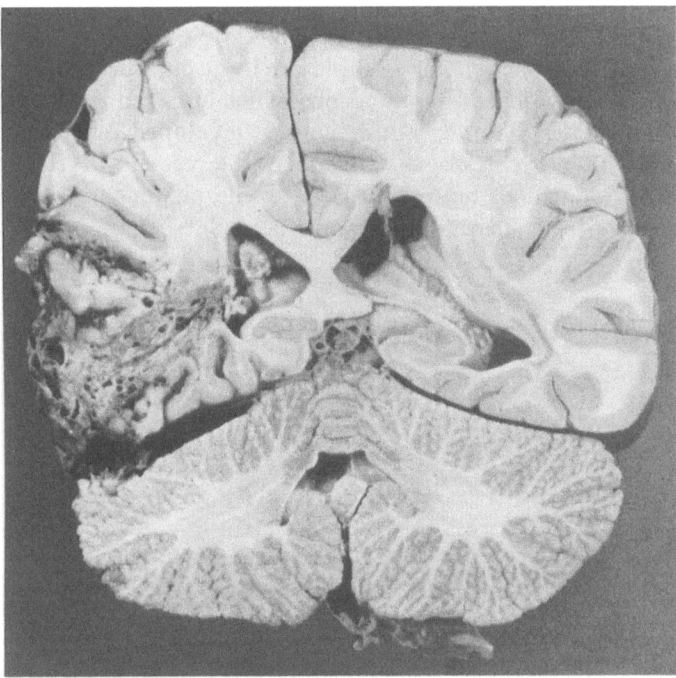

Fig. 7. An arteriovenous aneurysm of the middle cerebral artery in cross-section. E. H. A., 45 year old man 776/49. The patient had four episodes of intracranial bleeding prior to admission. Expired during the operativ procedure. The cuneiform lesion extends to the wall of the ventricle

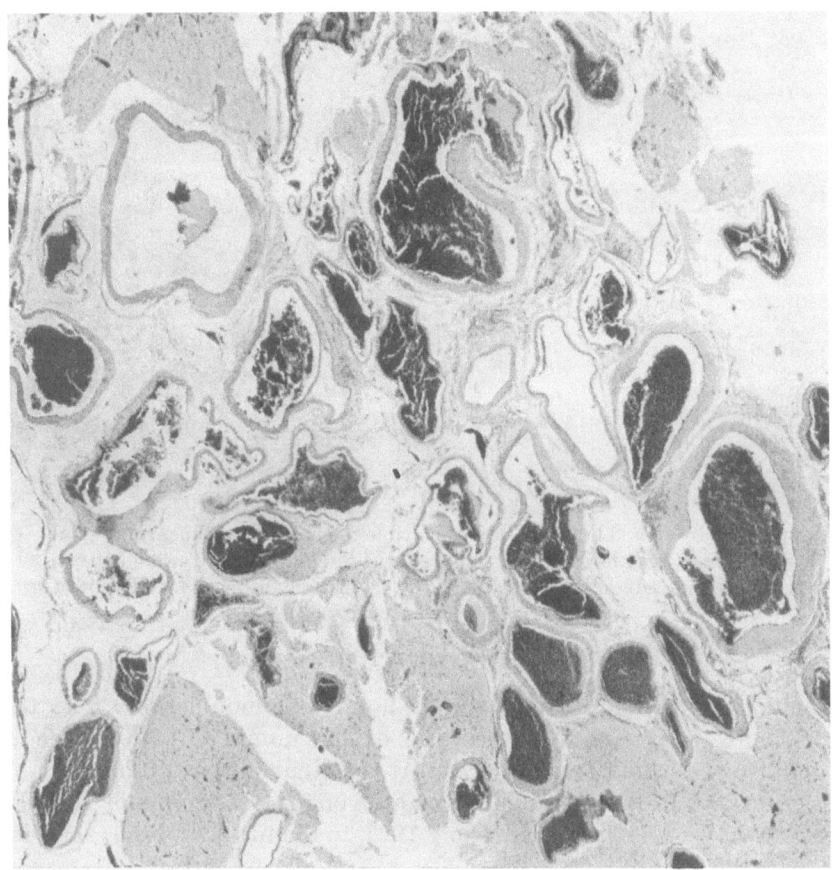

Fig. 8. Survey view. A differentiation between arterial and venous structure cannot be made. The irregular vessels with varying caliber are separated by glial tufts. The muscle layer is virtually absent (Dr. H. NORDENSTAM)

muscle. Often the sparse muscle layer is replaced by a curious proliferation of circular muscle producing the leiomyomatous nodulations which engaged the interest of the pathologists of the VIRCHOW era. The *adventitia*, poorly developed even for cerebral vessels, is especially deficient in adipose tissue. Between the vessels of the aneurysmal bed, cerebral tissue can be identified, however sparse, while evidence of gliosis and calcification attests to reaction of the surrounding brain tissues[1, 3, 7, 53].

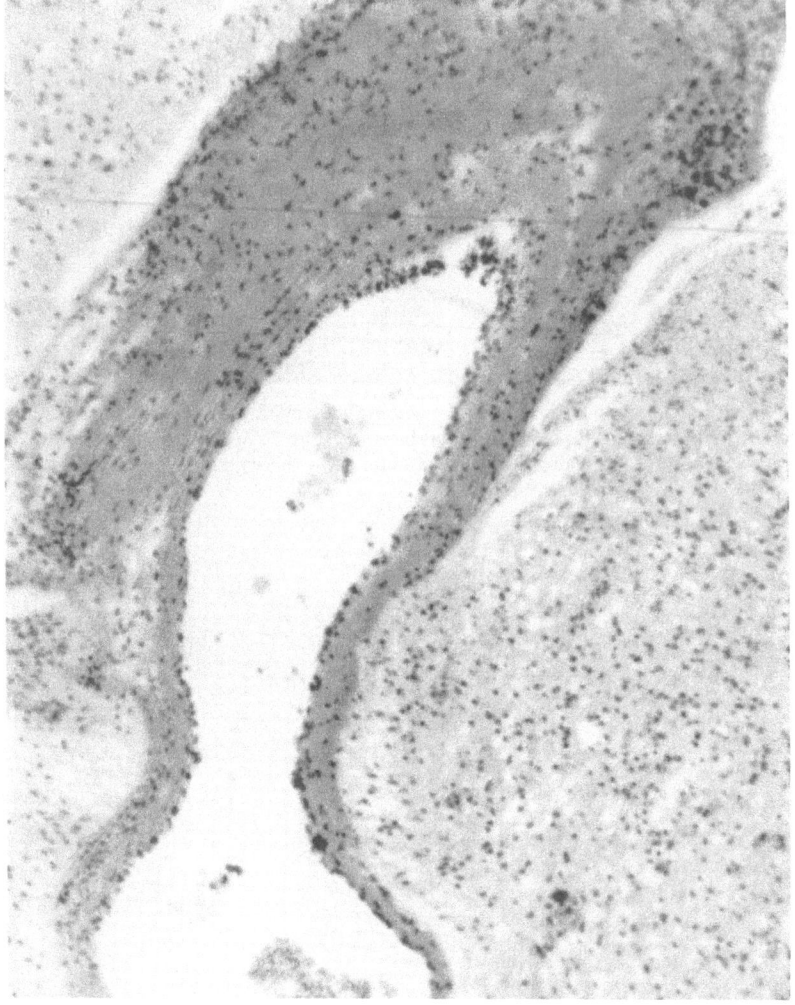

Fig. 9. A vessel surrounded by gliosed brain parenchyma. Note the varying thickness of the wall. The presence of inflamatory cells, as seen in the above slide, was a frequent finding not only around the vessels, but often in the vessel walls themselves. The muscular layer is virtually absent (Dr. H. NORDENSTAM)

# Incidence

The congenital arteriovenous aneurysm is not a common lesion. The traditional, albeit obscure, index of reference is the ratio of the number of arteriovenous lesions to the total number of verified brain tumors. At the Neurosurgical Clinic, Serafimerlasarettet, 5000 verified brain tumors and 125 arteriovenous aneurysms of the carotid and vertebral systems have been recorded from 1923 to January 1, 1955, an "incidence" of approximately 2%.

The congenital lesion occurred more frequently in the male. In this series 83 malformations occurred in the male sex and 42 in the female, an almost 2:1 ratio. Moreover, not only did the male predominate among the patients presenting with subarachnoid hemorrhage, as might be expected because of the greater stresses to which he is subjected, but an equal preeminance of the male sex was noted among the epileptic patients, and in every symptom complex and age grouping (Table 2, 3). This suggestion of predilection of the male is rather curious, in view of the equal proportion of sexes recorded in MACKENSIE'S series[33]. DANDY[8], in reporting his series of 8 male patients, suggested that with the accumulation of additional material, this possibly fortuitous proportion might be revised. Such has not occurred in this series to date, and a preponderance of males has been consistently encountered both in our own material and in the series of TÖNNIS and LANGE-COSACK[63]. The possibility that these statistics are within the pattern of chance cannot altogether be discounted, and it may be well to defer speculation until the next hundred cases have been accumulated in the literature.

Table 2. *Onset of the Presenting Symptom*

| Age | Intracranial bleeding | | | Epilepsy | | | Mild symptoms | | | Total | | |
|---|---|---|---|---|---|---|---|---|---|---|---|---|
| | M | F | T | M | F | T | M | F | T | M | F | T |
| 1–10 | 1 | 0 | 1 | 3 | 1 | 4 | 1 | 0 | 1 | 6 | 1 | 7 |
| 11–20 | 7 | 4 | 11 | 14 | 4 | 18 | 5 | 5 | 10 | 26 | 13 | 39 |
| 21–30 | 13 | 9 | 22 | 9 | 2 | 11 | 4 | 3 | 7 | 26 | 14 | 40 |
| 31–40 | 4 | 5 | 9 | 8 | 4 | 12 | 1 | 1 | 2 | 13 | 10 | 23 |
| 41–50 | 3 | 1 | 4 | 4 | 1 | 5 | 3 | 2 | 5 | 10 | 4 | 14 |
| 51–60 | 1 | 0 | 1 | 0 | 0 | 0 | 1 | 0 | 1 | 2 | 0 | 2 |
| Total | 29 | 19 | 48 | 38 | 12 | 50 | 16 | 11 | 27 | 83 | 42 | 125 |

Table 3. *Onset of Epilepsy and Intracranial Bleeding*

| Age | Intracranial bleeding | | | Epilepsy | | | Total | | |
|---|---|---|---|---|---|---|---|---|---|
| | M | F | T | M | F | T | M | F | T |
| 1–10 | 1 | 0 | 1 | 3 | 1 | 4 | 4 | 1 | 5 |
| 11–20 | 10 | 7 | 17 | 15 | 7 | 22 | 25 | 14 | 39 |
| 21–30 | 16 | 11 | 27 | 12 | 8 | 20 | 28 | 19 | 47 |
| 31–40 | 6 | 6 | 12 | 12 | 7 | 19 | 18 | 13 | 31 |
| 41–50 | 4 | 2 | 6 | 5 | 3 | 8 | 9 | 5 | 14 |
| 51–60 | 0 | 0 | 0 | 2 | 0 | 2 | 2 | 0 | 2 |
| Total | 37 | 26 | 63 | 49 | 26 | 75 | 86 | 52 | 138 |

One might predict that in arteriovenous aneurysms of congenital origin, the first symptom would appear at an early age. The earliest onset of symptoms (epilepsy) encounted in this series occurred at the age of 4, and although symptoms have been reported at birth[11], and at 8 weeks[26], 8 months[4], $2^1/_2$ years[44] and 3 years[68] of age, the incidence of lesions presenting in the first 10 years of life in this series was less than 5%. In the majority of our patients the first symptoms developed in the second or third decade, about the same time as arteriovenous aneurysms of the extremities are said to become manifest[6, 28, 57, 67]. This tardy appearance of symptomatology has correspondence with the slow sequence of morphologic change.

# Location

## A. External Carotid Lesions of the Extracerebral Structures

Lesions of the external carotid artery situated in the more superficial structures do not have remarkable potentialities for derangement of the circulatory system nor do they result in intracranial hemorrhage. Nevertheless, the need for surgical intervention is at times considerable. *1. Exsanguination.* In the normal scalp, the interlacing fibrous network prevents a retraction of the lacerated blood vessel, so that the loss of blood from even a small wound can at times be so great as to constitute a threat to life. The obvious hazard in this age of violence from the presence of a huge mass of engorged, elastin-deficient vessels in an exposed area of the body requires no elaboration (Fig. 10). Control of bleeding from a wound of these vessels is unquestionably beyond the scope of the physician in the emergency room, and long before a diagnosis can be established, the patient may die from hemorrhage. *2. Bruit.* A subjective auditory complaint referred to the head is an almost invariable part of the symptom-complex, especially irritating to the patient when he rests his head against a pillow. Some describe the noise as booming, others liken it to a rhythmic tattoo on the head with a hammer. Although compatible with life, this exasperating symptom has a most distressing effect on the patient's psychic resources.

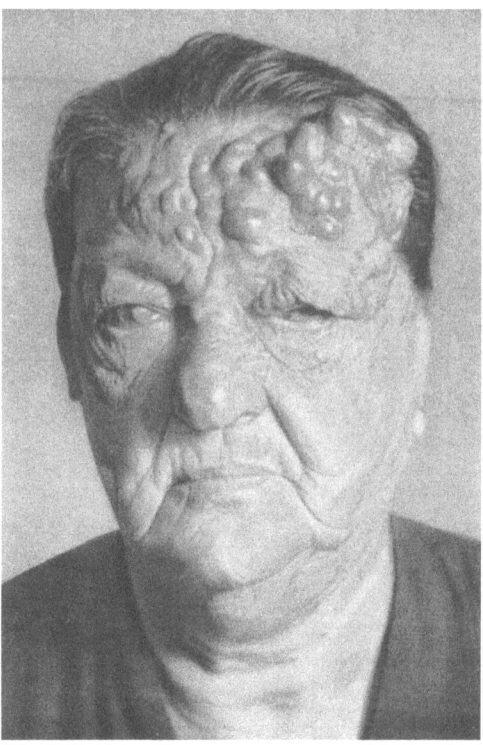

Fig. 10. An arteriovenous aneurysm of the external carotid artery (Case of Prof. PAULO NIEMEYER, Rio de Janeiro)

In comparison with malformations of the brain, extracerebral lesions of the external carotid artery are even more unusual, occurring at a ratio of about 1 to 20. In the era preceding the advent of angiography, this relationship was obscured by difficulties in the diagnosis of the brain malformations. Lesions reported in the 19th century were invariably situated in the superficial circulation, for here the presence of dilated, pulsating vessels, bruit and thrill did not overtax the diagnostic resources of the examiner. Not until the advent of the contrast study was the comparative rarity of extracranial lesions of the external carotid artery demonstrable.

In the present section we will review those lesions of the external carotid artery situated in the superficial tissues and dura, in which all, or almost all, of the blood supply demonstrable by angiography came from the external carotid artery. We will exclude from discussion the external carotid lesions located within the brain substance; lesions in which the external carotid contributed only a small share of the total arterial supply; and those lesions within the cerebral substance where, for some unknown reason, the external carotid was dilated, although no direct communication with a deep lesion could be demonstrated.

The frequent occurrence of secondary involvement of the external carotid artery having been excluded from discussion, a total of 7 extracerebral lesions can be listed, 6 of which were fed exclusively by branches of the external carotid artery, while one contained in addition a small contribution from the deeper circulation.

When the malformation is situated between galea and skin, not only does the vascularity of the scalp add to the difficulties of excision but one must also be prepared to resort to skin graft, if large defects remain after the removal of the involved epidermis. The interposition of skeletal muscle in the nuchal and face regions presents an added problem. Before considering the extirpation of malformations of the dura, the surgeon must not be misled in the belief that these lesions are invariably supplied by meningeal branches of the external carotid artery. Instances have been reported where the dural lesion has been supplied by the internal carotid artery[51].

Since in the section on treatment the remarks will chiefly concern lesions of the brain, it might be well at this time to relate in some detail our experiences with these extra-cerebral lesions.

## 1. Frontotemporal Region

Inasmuch as the malformation commences early in Stage 2, it can be appreciated that if transitory communications between internal and carotid circulations were to appear in Stage 4, the stimulus of increased pressure within the external carotid artery might prevent the normal course of obliteration. Such temporary communications do,

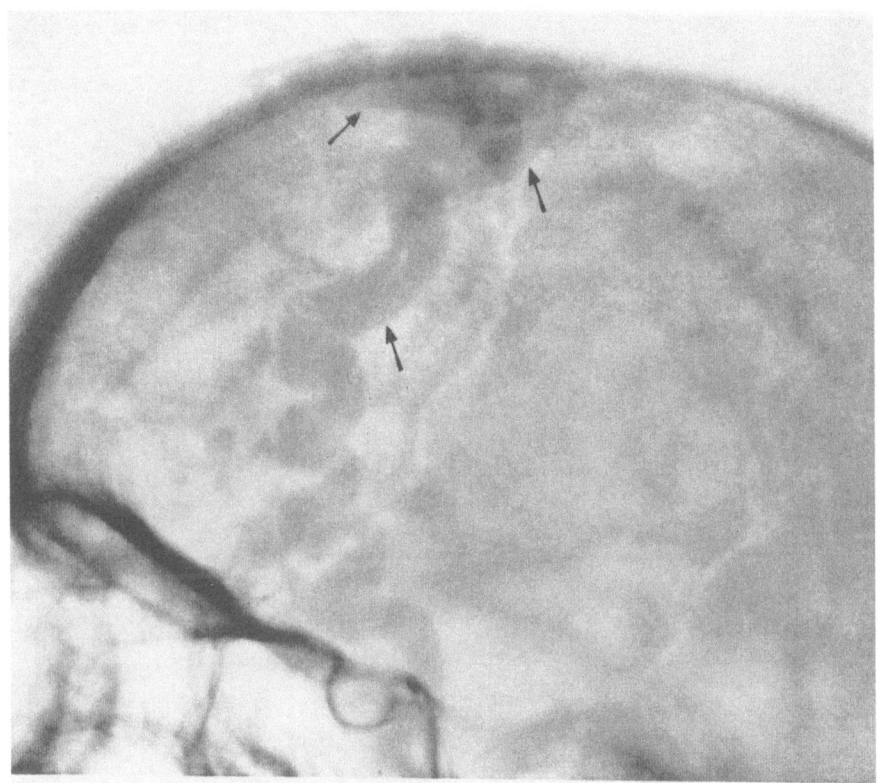

Fig. 11. Case I. *Right external carotid angiography* (preoperative). An extracranial lesion situated chiefly to the right of the middle. Supplied by the superficial temporal artery and a branch from the meningeal artery. Part of the aneurysm is situated in the bone. Drainage by a vein coursing in front of the coronal suture and by another passing down the middle of the forehead

in fact, occur during embryonic development, and their persistence has been noted in one patient with a lesion in the frontotemporal region.

In the 6 week human embryo several primitive ophtalmic branches of the internal carotid artery provide the early vascular network around the ophtalmic stalk. The primordial hyoid-stapedial artery arises from the internal carotid artery, the stapedial division of which courses from the petrous region to the orbit. When obliteration occurs at the junction of the hyoid-stapedial and internal carotid arteries, circulation is routed through the meningeal branch of the external carotid artery. With the development of the definitive ophtalmic artery which arises from the internal carotid artery, the communication between the external carotid and the stapedial arteries also becomes oblite-rated. The terminal branches of the stapedial artery persist as the supraorbital artery, but are absorbed into the internal carotid system[61, 62].

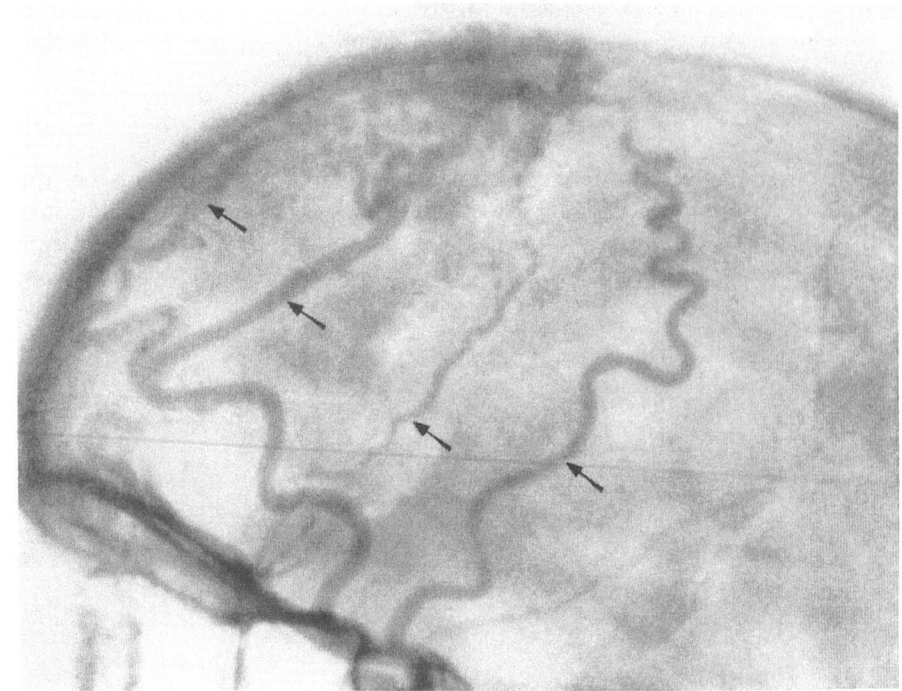

Fig. 12. See text of Fig. 11

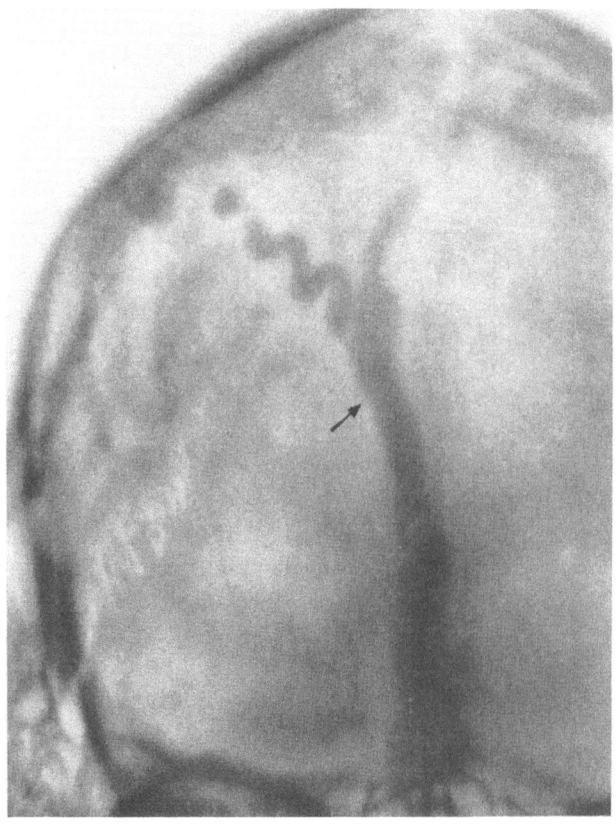

### Case I

Woman, age 20. External carotid lesion of the frontotemporal region. Extirpated after three surgical procedures. Good result.

*First admission.* (Oct. 29, 1951). E.N., 841/51. 20 year old woman who, for 8 months prior to admission, noted a pulsating swelling in the right frontotemporal region which first appeared during pregnancy, subsequently becoming larger. *Examination* revealed a pulsating frontotemporal swelling, associated with a loud murmur which was intensified during systole. *Skull:* increased vascularity

Fig. 13. See text of Fig. 11

in the right frontal region. *Angiography* (Fig. 11–14): a large arteriovenous aneurysm in the scalp and in the bone in the right frontal region. On the left side, although some dilatation of the branches

of the external carotid artery was present, the changes were less pronounced than on the right. The right middle meningeal artery was moderately enlarged. Arteriograms of both internal carotid arteries demonstrated an enlargement of both ophtalmic arteries with free communication with the aneurysm. *First operation* (Nov. 15, 1951). After preliminary ligation of the right external carotid artery, a large skin-galea flap was turned back in the right frontotemporal region. Numerous arterialized veins perforating the galea required clipping or coagulation during the dissection of the flap. Beneath the periosteum a tangle of dilated blood vessels was encountered and the operator took this to be the main lesion. After a bone flap was turned back, the middle meningeal artery was found to be markedly dilated. This vessel was divided. The dura was then opened, the flap completely devascularized and sutured in place. Thereafter, the bone flap was freed from its attachment to the temporal muscle and wired to adjacent bone. *Postoperative course.* The wound healed, the swelling and murmur disappeared and the patient was discharged. The patient remained symptom-free for almost a year but during her second pregnancy she noted the return of the systolic murmur and a progressively enlarging frontal swelling (Fig. 15).

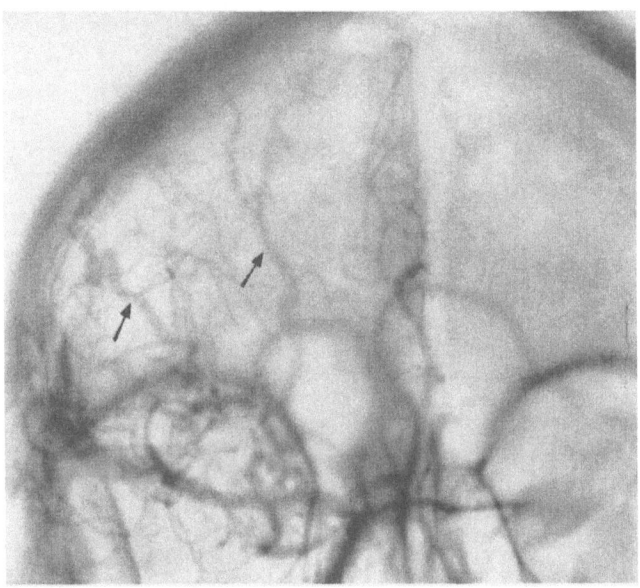

Fig. 14. Case I. *Right internal carotid angiogram* (preoperative). The opthalmic artery is dilated. Many tortuous vessels are seen in the orbit. The supraorbital branches feeding the malformation are supplied by the opthalmic artery

*Second admission* (Feb. 22, 1954). *Angiography* (Fig. 16, 17): the aneurysm was localized to the scalp, fed by branches of both external carotid arteries with some contribution from both ophtalmic arteries. All of the branches of the external carotid except the middle meningeal were enormously dilated. *Second operation* (March 3, 1954). A large coronal incision was extended from one zygoma to the other. After both temporal arteries had been ligated, the flap and periosteum were dissected down to the orbital ridge. The blood vessels entering the flap in this area were ligated. The aneurysm, situated between the galea and epidermis, was then dissected out and excised. In some areas of skin, where the lesion was intimately fixed to epidermis a few minor defects in the skin resulted from excision of the lesion. *Third operation* (March 18, 1954). As a precautionary measure, both occipital arteries were ligated since, following the second procedure, they appeared to be dilated. *Postoperative course.* The wounds healed without complication. The patient has been symptom-free for almost 2 years.

**Comment.** Since the introduction of hypotension anesthesia, the necessity for exposure of the external carotid artery in the neck has largely been obviated.

A lesion of similar location was observed in a 10 year old boy (J. B. E., 901/53), who had a lump on his forehead for 3 years before admission. Angiography demonstrated

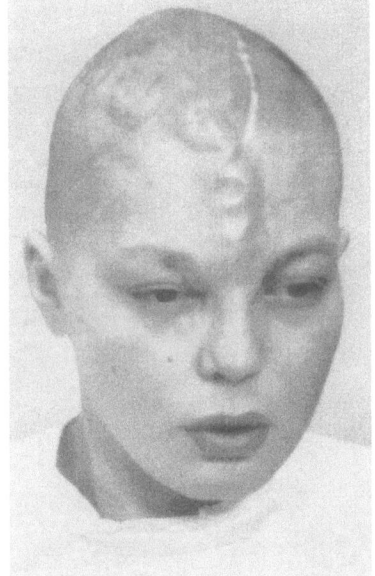

Fig. 15.
Case I. Prior to the second operation

that the lesion was fed by a frontal branch of the external temporal artery without involvement of the deep circulation (Fig. 18). The tumor was excised without difficulty. There has been no recurrence.

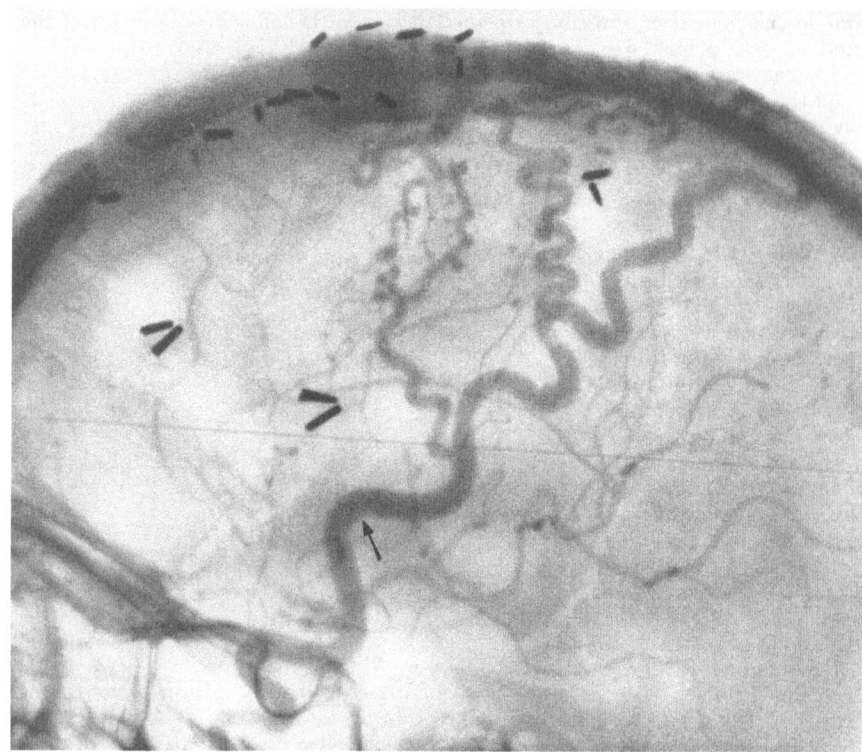

Fig. 16. Case I. *Right common carotid angiography* (2nd admission). The residual lesion is situated in the galea and extends from the nasion to above the lambda. Located chiefly on the right side. The lesion is supplied by the branches of the superficial temporal artery and the opthalmic artery

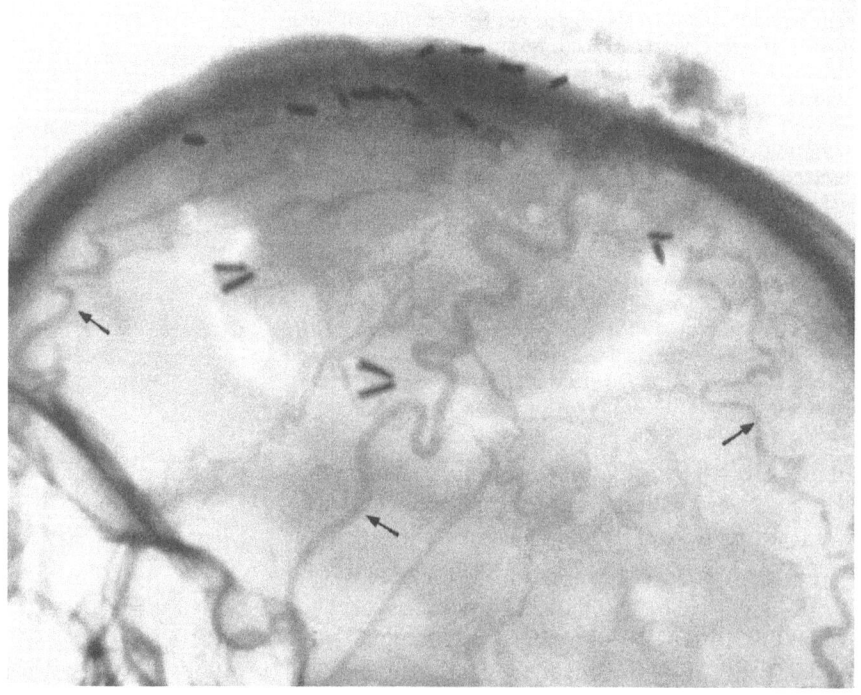

Fig. 17. Case I. *Left external carotid angiogram* (2nd admission). Contributions from the frontal, temporal, occipital and middle meningeal arteries

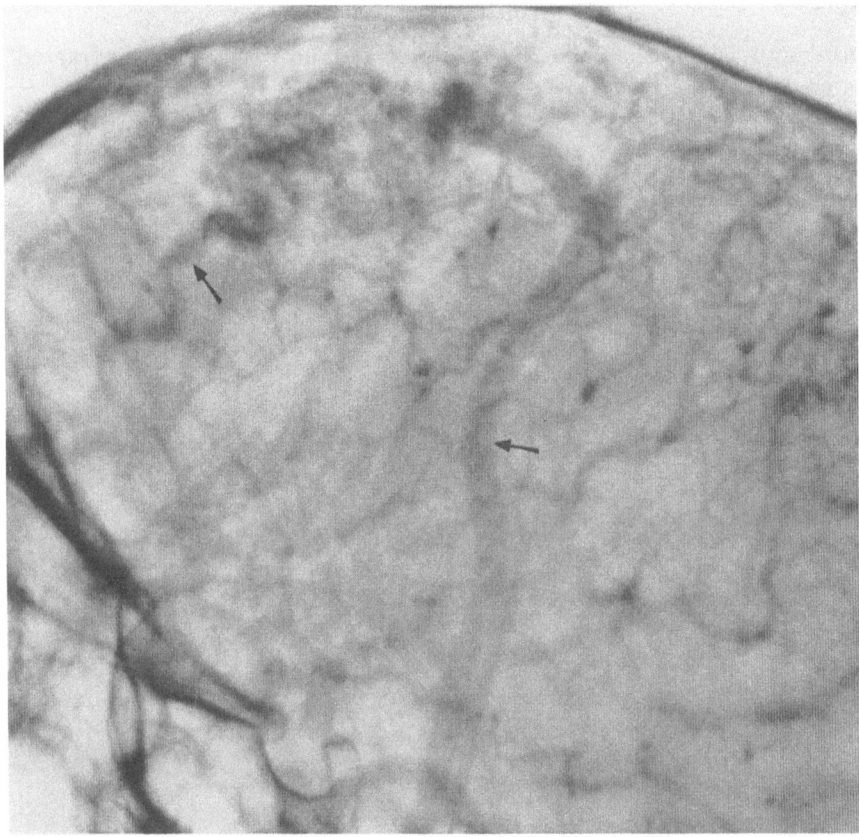

Fig. 18. J. B. E., 10 year old boy, 901/53. *Right common carotid angiogram.* A dilated frontal branch of the superficial temporal artery supplies the lesion. Drainage by the dilated temporal vein. The aneurysm is extra-cranial

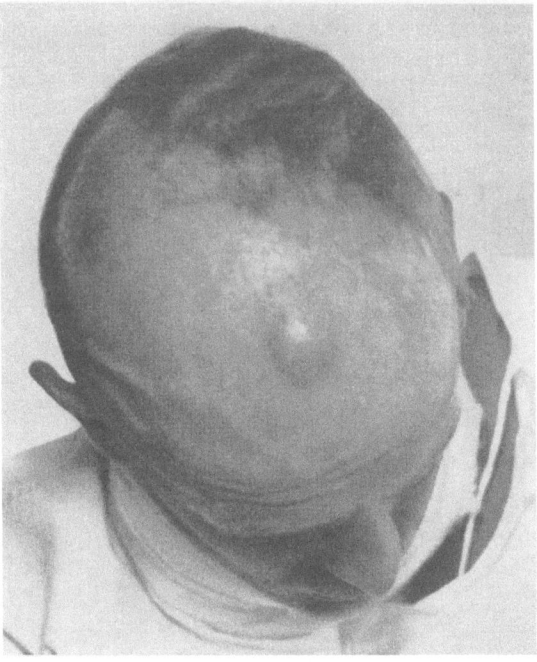

Fig. 19. Case II. Lesion of the right temporal branch of the external carotid artery

## 2. Frontoparietal Region

The ease with which the following lesion could be excised is contrasted with the difficulties encountered in case I.

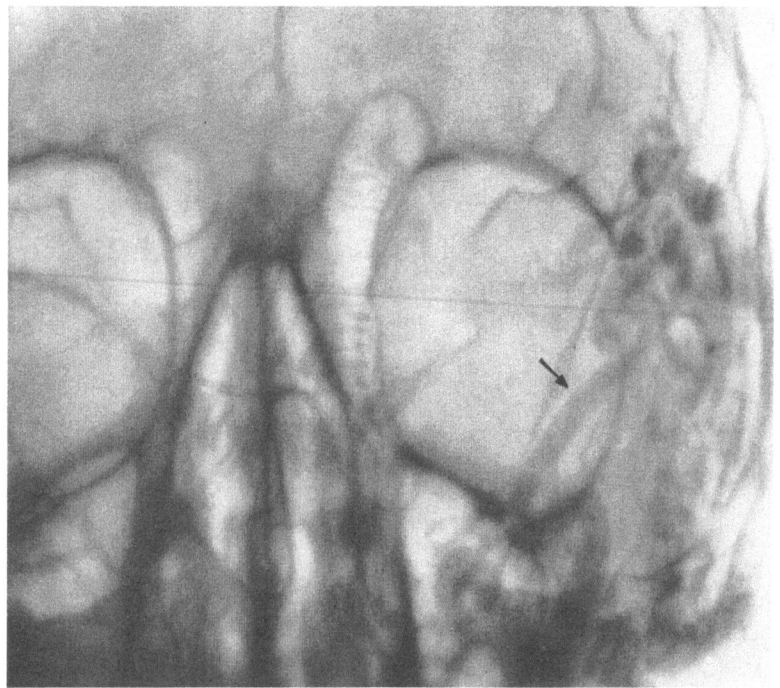

Fig. 20. P. G. P., 39 year old man, 612/47. *Left external carotid angiogram* (preoperative). Lesion situated in the anterior part of the temporal region, supplied by the internal maxillary artery. Drainage to the deep facial vein

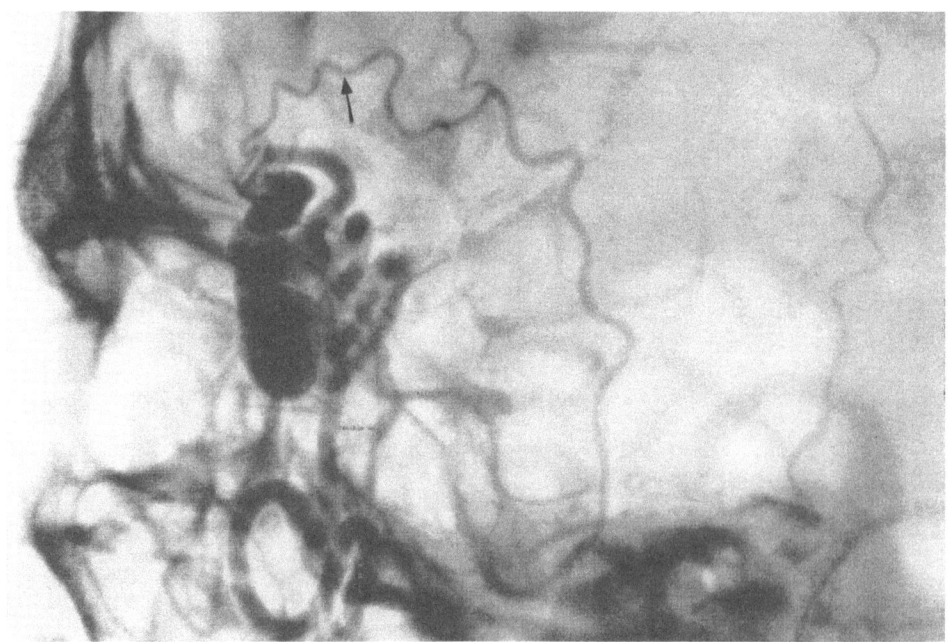

Fig. 21. See text of Fig. 20

## Case II

Man, age 28. Lesion of the frontoparietal region, supplied by the right temporal artery. Extirpated. Gainfully employed.

*Admission* (June 16, 1936). K. R. W., 343/36, a 28 year old man, had noted a slowly enlarging, pulsating lump on his scalp for $1^1/_2$ years, unassociated with subjective auditory disturbance (Fig. 19). *Arteriography* of the right external carotid revealed an arteriovenous aneurysm fed by branches of the right temporal artery. *Operation* (June 28, 1936). After ligation of the right temporal artery, the mass ceased to pulsate and could be excised without difficulty. *Clinical course*. The patient returned to his previous occupation. There has been no recurrence of the lesion.

**Comment.** This is the only instance in this series where a bruit or subjective murmur was absent in a lesion of the external carotid artery. Had the lesion been of longer duration, it would seem likely that a murmur would eventually have appeared.

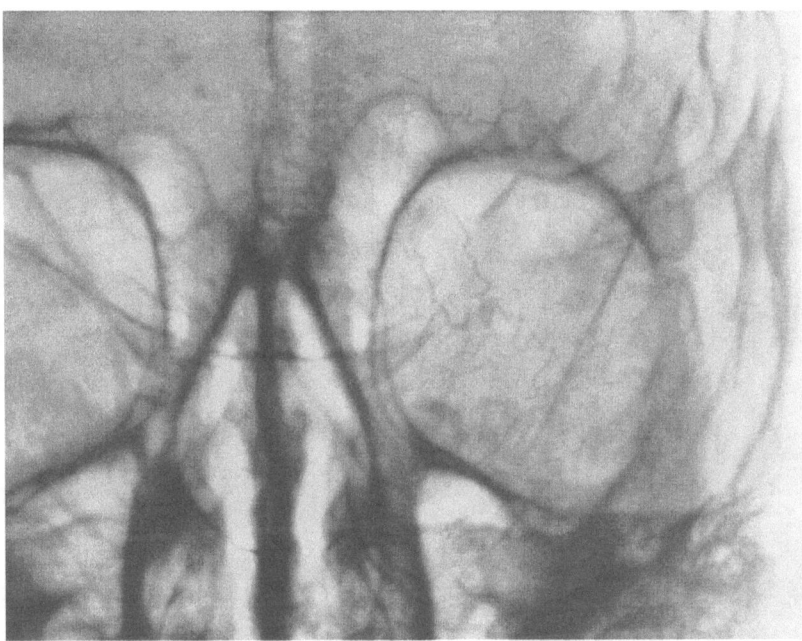

Fig. 22. P. G. P., 39 year old man, 612/47. *Left external carotid angiogram* (postoperative). The lesion has been removed

## 3. Temporal Region

In the temporal area a capsule-like enclosure formed by the adjacent tissue frequently circumscribes the malformation, permitting a relatively easy removal of the lesion. Beginning at the periphery of the lesion, the excision is extended down to bone. All the tissue enclosing the aneurysm should be dissected without disturbing the interior of the aneurysmal mass. Such a lesion, encountered in a 39 year old male (P. G. P., 612/47), was situated between galea and epidermis. The lesion was excised without difficulty and there has been no recurrence (Fig. 20–22).

## 4. Nuchal Region

The interposition of skeletal musculature in the nuchal area transforms the somewhat circumscribed vascular mass encountered in the temporal region into a tangle of vessels situated within and without the muscle bundles. In our fourth patient (L. M., 395/43) two unsuccessful attempts were made to remove the mass of blood vessels, the procedures accomplishing little more than the ligation of some of the afferent arteries. Only when the aneurysm was dissected out from its bed within the deep muscles during the third attempt was a satisfactory result attained.

## 5. Face

The presence of skeletal musculature in the face similarly increases the difficulty of extirpating a lesion situated in this region. The removal of large masses of muscle may be necessary before a good result can be anticipated. Great care must be taken to preserve the branches of the facial nerve.

### Case III

Man, age 32. Arteriovenous aneurysm of the face, fed by the external carotid artery. Extirpated after two surgical procedures. No recurrence[13].

*First admission* (Dec. 4, 1939). A. G. L., 454/39, a 32 year old man with a history of a tumor slowly enlarging on the left side of the face since childhood. *Examination* revealed a deep, pulsating

Fig. 23. Case III. Lesion of the external carotid artery in the face

Fig. 24. Case III. The lesion has been removed. This retouched photograph has been submitted by the patient

mass the size of a goose egg, over which a murmur could be auscultated (Fig. 23). *Angiography* of the left external carotid artery disclosed a mass of dilated blood vessels behind and above the external maxillary artery and dilatation of the external carotid system. *First operation* (July 12, 1939). The skin incision extended from the lobule of the left ear over the anterior border of the tumor to the left ala nasi. After reflection of the skin flap, two branches of the facial nerve were identified. The aneurysm was found medial and deep to the parotid gland. The parotid could be separated from the zygoma, with preservation of the facial nerve branches. The upper part of the parotid gland was separated from the masseter muscle in which the bulk of the aneurysm was situated. After the arterialized vessels had been ligated and divided, both angiomatous tissue and masseter muscle were removed *en bloc*. Unfortunately a part of the lesion situated posteromedially in the pterygo-palatine fossa was not easily accessible and the operation had to be terminated before complete removal could be effected. *Pathological report:* Intramuscular racemose angioma. *Postoperative course.* The swelling recurred soon after surgery and the patient left the clinic unimproved.

*Second admission* (Dec. 13, 1943). *Second operation* (Dec. 16, 1943). Local anesthesia. The incision extended along the posterior border of the left mandible and another incision was made below the lobe of the ear in a direction extending backward to join the main incision. A third incision was made in the direction of the wound of the first operation so that the final pattern was cruciate in shape. The branches of the facial nerve were identified. The parotid capsule was incised and the gland mobilized anteriorly and from below. A bundle of arterialized veins about the size of a small orange was seen between the parotid gland and the remnants of the masseter muscle. Afferent and efferent branches, including a large arterialized vein which emptied into the facial vein, were systematically ligated and the remaining portion of the masseter muscle was removed. The aneurysm extended into the pterygopalatine fossa, and after removal of the accessible vascular tissue, it appeared that a stump of blood vessel about the size of a thumb tip remained behind in the fossa. Branches from the internal maxillary artery were clipped and divided. Facial nerve branches were stimulated and found to be intact. Closure with drain. *Postoperative course.* A transient facial nerve paralysis persisted for several months before the return of complete function. No recurrence of the lesion (Fig. 24) in 15 years.

**Comment.** The incisions by present-day head and neck surgery standards were certainly crude, but it must be added that this is perhaps the first successful removal of an arteriovenous lesion deep to the parotid gland with preservation of the facial nerve.

## 6. Dura

Location of the arteriovenous aneurysm in the dura can be troublesome. Not only must the surgeon insure that his incision is sufficiently large to mobilize all the feeding branches, but he must be prepared to deal with large dural defects. In our seventh patient (K. R. W., 343/38), the malformation was located on dura, fed by branches from the middle meningeal and occipital arteries. Bleeding could not be controlled by plugging the foramen spinosum. Reoperation through a larger exposure was necessary before the contributing branches from the occipital artery could be effectively ligated and the mass extirpated along with the dural attachment.

## B. External Carotid Lesions of the Brain

Five cases occurred in this series in which a lesion situated on or within the brain substance was fed predominately or exclusively by the external carotid artery. In a previous section it was noted that the interposition of the meninges and cranium may incompletely exclude the external carotid network from the intracerebral circulation. We will now indicate that not only can a direct connection between the two systems exist, but a large area of the brain surface may derive its arterial supply predominately from the external carotid artery (Fig. 2,*o*). Accordingly, these cerebral lesions of the external carotid artery often present great difficulties, for they combine the problems of both superficial and deep arteriovenous aneurysms.

### 1. Frontal Lobe

The following case history will relate a most important lesson derived from the experience of this clinic with arteriovenous aneurysms.

### Case IV

Woman, age 44. Frontal lobe lesion fed by external carotid and internal carotid branches. Operative death[40].

*Admission* (July 9, 1939). E. S. A., 467/39, a 44 year old woman with a 15 year history of increasing vascularity of the scalp in the form of a nest of dilated, pulsating blood vessels. A bruit had also been noted by the patient. *Angiography* of the external carotid revealed an arteriovenous aneurysm fed by an enormously dilated middle meningeal artery (Fig. 25–27). *Operation* (July 19, 1939). The lesion was belatedly observed to extend deep into the frontal lobe with contributions from the internal carotid artery. Severe hemorrhage was encountered in the course of the exploration and the patient expired. *Postmortem* disclosed a typical wedge-shaped aneurysm involving a large part of the left frontal lobe (Fig. 28).

**Comment.** The disastrous course of this operation was entirely attributable to the omission of a preoperative internal carotid angiogram. A lesion of the external carotid artery *always* requires a survey of the ipsilateral internal carotid and the contralateral external carotid systems.

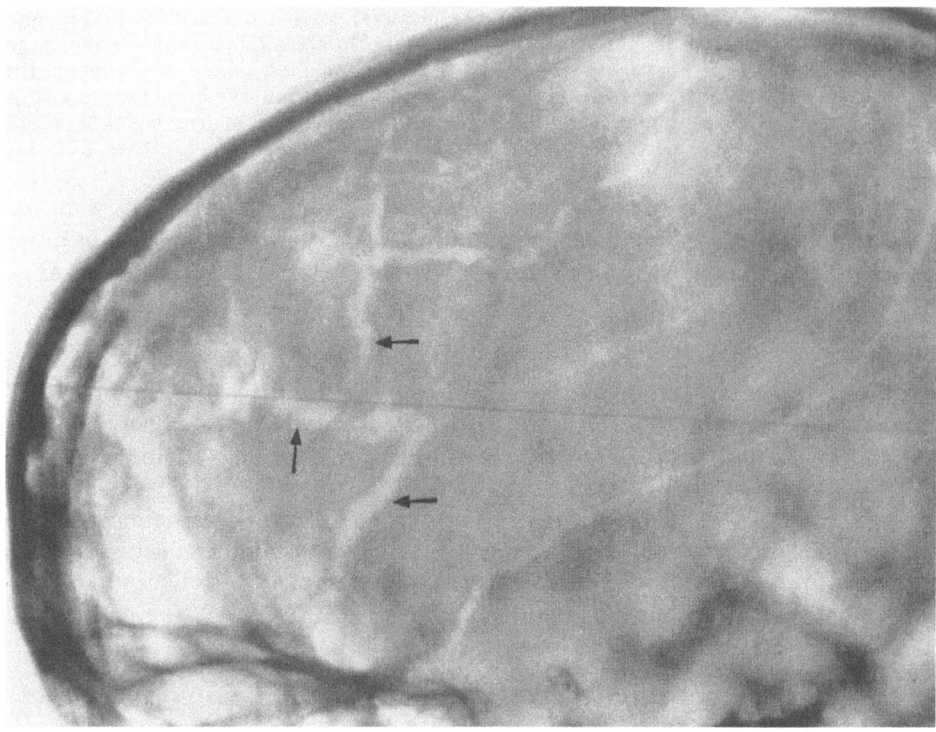

Fig. 25. Case IV. *Skull films.* Irregular grooving in the left frontal region which continues down to the base of the skull. The groove of the sagittal sinus is also enlarged. (Enlargement of the foramen spinosum on the left can be seen in other views)

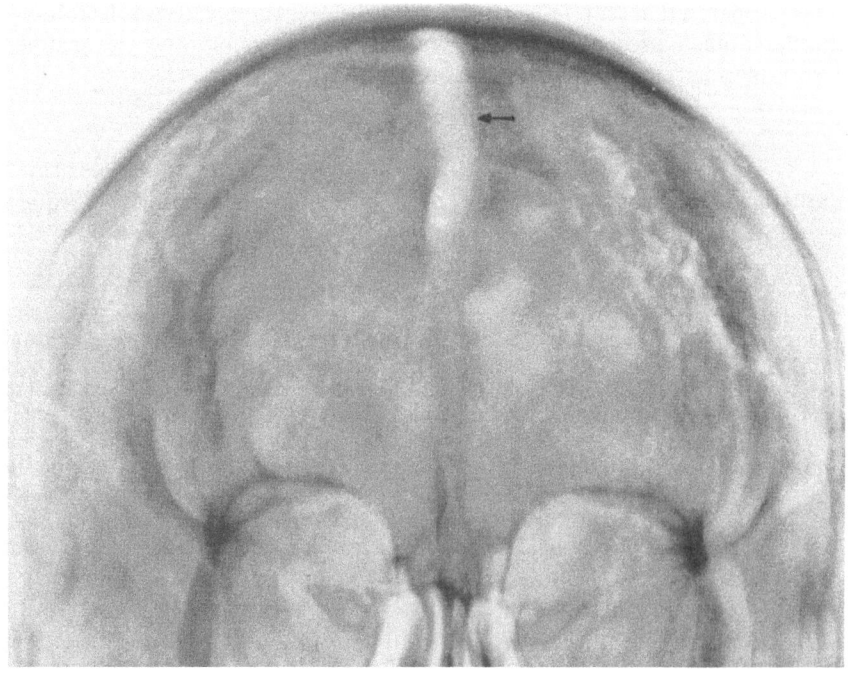

Fig. 26. See text of Fig. 25

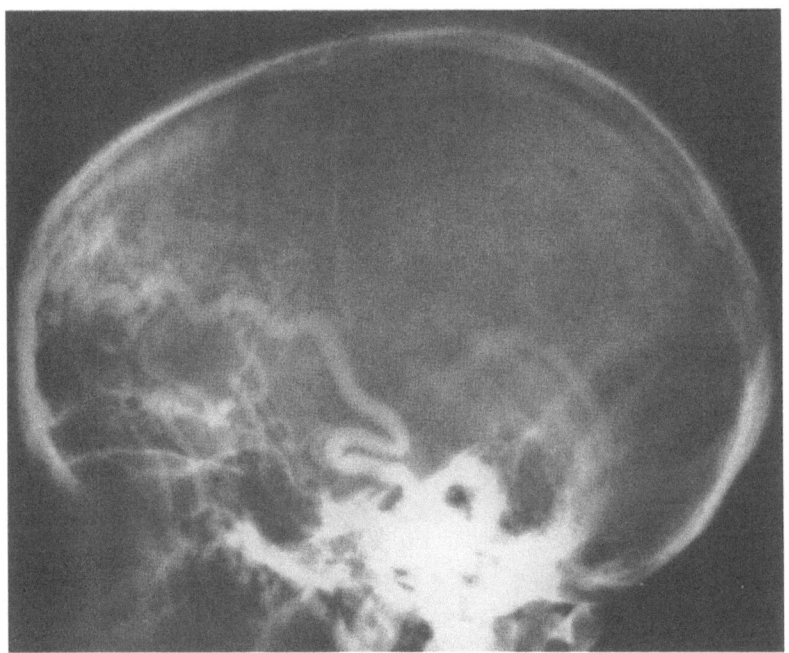

Fig. 27. Case IV. *Left external carotid angiogram.* The external carotid artery is dilated. In the left frontal region a branch of the middle meningeal artery and a branch of the internal maxillary artery supply the aneurysm. Drainage to the superior sagittal sinus

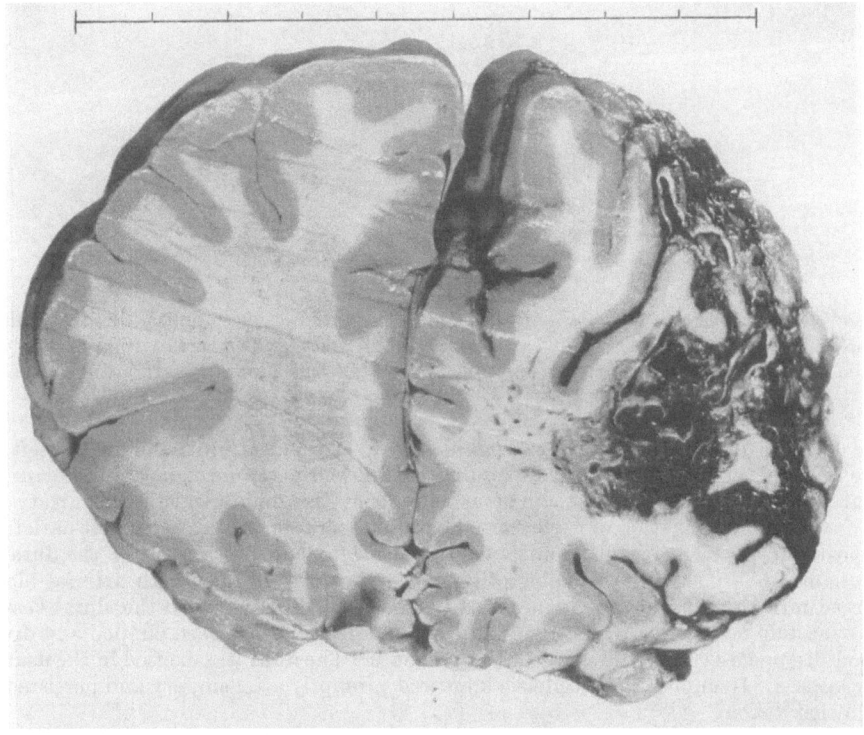

Fig. 28. Case IV. Autopsy section. A cuneiform lesion within the brain substance. Clinically, the malformation presented as a lesion of the external carotid artery

## 2. Parietal Lobe

Lesions of the external carotid artery situated in the parietal area must be treated with the same deference as a lesion in the same area supplied by the deep circulation, for the sources of arterial blood supply to the surface of the brain may be uncertain.

### Case V

Male, age 53. External carotid lesion of the parietal lobe. Excision. Hemiplegia and aphasia[3].

*Admission* (May 22, 1933). K. A. O., 1311/33, a 53 year old farmer with a 2 year history of attacks of syncope and speech disorder and a 6 months history of a bruit located in the left temporal region,

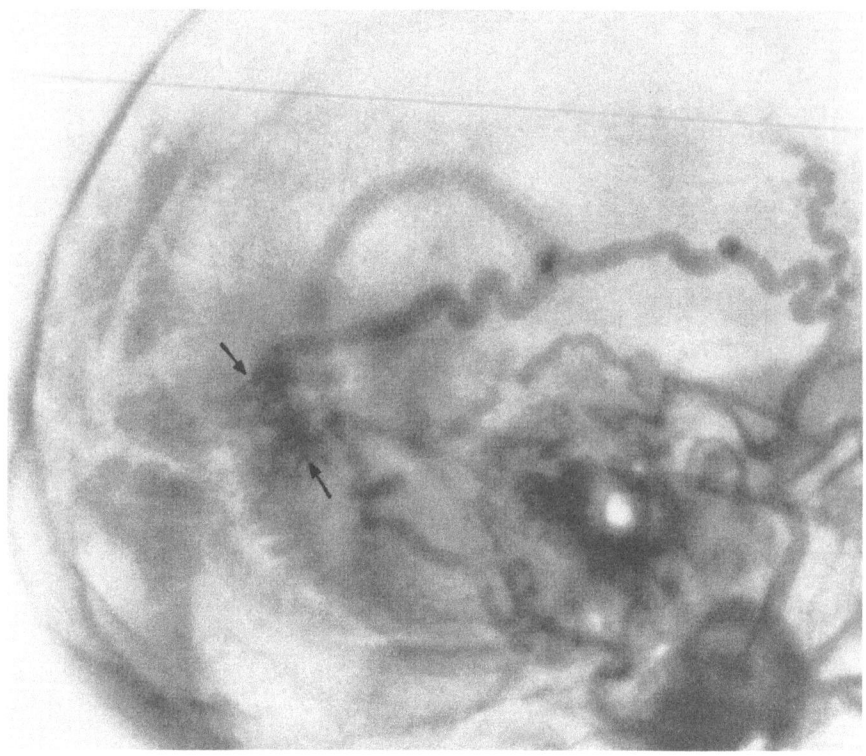

Fig. 29. Case V. *Left common carotid angiography.* A large cerebral lesion supplied chiefly by the anterior and posterior branches of the middle meningeal artery, the external maxillary artery, and probably a branch from the posterior cerebral artery. Inion lies to the left and auditory meatus to the right

accentuated before a fainting attack. *Examination* revealed a pulsating tumor in the left temporal region. The remainder of the neurological examination showed no abnormality. *Angiography* of the left external carotid disclosed dilated, tortuous masses of the middle meningeal artery (Fig. 29). *Operation* (May 30, 1933). Ligation of the external carotid artery on the left side. Left parietotemporal flap. The posterior branches of the middle meningeal artery and the vessels of the dura were tortuous and dilated. The pulsating pial veins of the parietal area were filled with arterial blood which was discharged into the veins of the pia mater at three sites of junction with the dural vessels in the region of the mastoid emissary. After these three communications had been clipped and divided, the arterial blood disappeared from the veins of the pia mater. The dura was excised in the usual fashion. *Postoperative course.* Hemiplegia and aphasia appeared promptly after surgery and persisted throughout the following years.

**Comment.** The three small communications between the dural lesion and the veins of the pia mater could be excised with ease and without the slightest discernable gross trauma to the brain. It is apparent, therefore, that a large part of the circulation of the brain was derived from the external carotid artery.

### 3. Temporal Lobe

A classic example of a middle meningeal lesion, presenting as dilated scalp veins, occurred in a 47 year old woman, A. E. H., 884/45, who had noted a blowing noise in the right ear for many years. A murmur and thrill were present on examination. Angiography revealed a large lesion of the temporal lobe, fed by the middle meningeal vessel with contributions from the posterior cerebral artery. Because of the paucity of symptoms and the hazards of surgery, operation was deferred. The patient died three years later, presumably from intracranial hemorrhage.

### 4. Occipital Lobe

These lesions are usually of a resectable nature and every effort should be made to remove them.

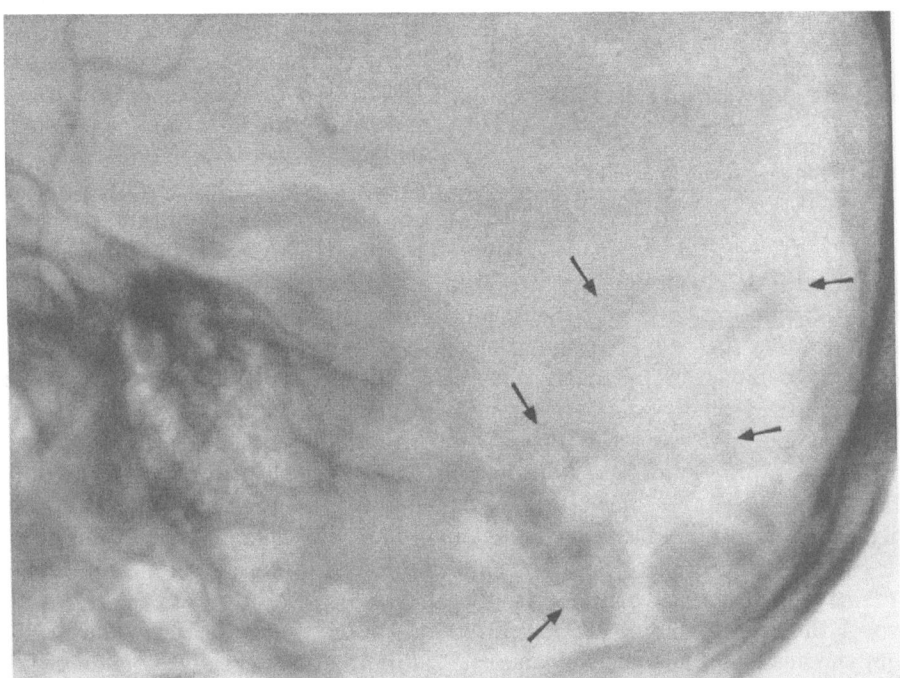

Fig. 30. Case VI. *Right external carotid angiography.* A lesion of the occipital lobe supplied by the external carotid artery. Right internal carotid angiography failed to demonstrate filling of the lesion

### Case VI

Man, age 46. Lesion of the occipital pole. Resection. Gainfully employed.

*Admission* (May 2, 1946). S. A. S., 384/46, a 46 year old male, who after periodic attacks of scotoma for one year, developed subarachnoid bleeding. *Examination* disclosed a mild right facial paresis and nystagmus on left gaze. *Angiography* revealed an arteriovenous aneurysm located in the dura and extending into the occipital pole (Fig. 30). *Operation* (May 9, 1946). Right occipital flap. Evidence of an old intracranial clot was observed, the previous hemorrhage having destroyed a large part of the lesion. The remaining malformation was removed by block dissection and the dura excised without replacement. *Postoperative course.* Recovery was uneventful apart from one epileptic attack and the need for readjustment of the bone flap. The patient has since returned to his previous employment.

**Comment.** The intracranial bleeding occurred approximately six months before surgery and offered no impediment to operation. On the contrary, not only did the bleeding shorten the work of the surgeon by destroying a part of the lesion, but after the clot had been evacuated, the surgeon found the lesion partly mobilized.

Our fifth patient, a 52 year old woman (V. M. D., 544/38), had an almost identical lesion situated in the occipital pole, fed by branches of the middle meningeal artery.

At operation evidence of intracranial bleeding was encountered, which destroyed the bulk of the lesion. The remainder of the malformation could be resected without difficulty and the patient has since had no further trouble.

## C. Internal Carotid and Vertebral Arteries

The census of malformations in this series is presented in Table 4.

From the above it can be seen that more than half of all arteriovenous aneurysms situated within brain substance were supplied by the middle cerebral artery. Next most frequent were lesions of the anterior cerebral artery, and the least common were malformations receiving their supply solely from the posterior cerebral artery.

*Table 4*

| | |
|---|---|
| *Internal carotid artery* . . . . . . | 107 |
| Anterior cerebral . . . . . . . 24 | |
| Anterior + middle cerebral . . . 10 | |
| Middle cerebral . . . . . . . 64 | |
| Middle + posterior cerebral . . . 6 | |
| Posterior cerebral . . . . . . . 1 | |
| Posterior cerebral + vertebral . . 3 | |
| *Vertebral artery* . . . . . . . . . . | 6 |
| *External carotid artery* (superficial) . | 7 |
| *External carotid artery* (deep) . . . | 5 |
| *Total* . . . . . . . . . . . . . | 125 |

The lesions of the deep circulation are always located on and within brain substance. The afferent vessels entering the brain at some distance from the malformation divide into numerous small, thin-walled vessels which penetrate the lesion at a deeper level, thereby producing the characteristic cuneiform structure which frequently extends to the vicinity of the ventricle. Parenthetically, it can be seen from this consideration that an attempt to effect a material alteration of the size of the lesion by ligating or cauterizing the vessels of the surface is certain to prove ineffectual.

*Choroid artery.* Seven malformations in this series received some contributing branches from the choroid arteries. Angiography established the arterial supply in 5 instances, while in the remaining 2 cases the observation was made at the operation table. In the absence of angiographic evidence, however, gross appraisals can often be erroneous, for a vessel containing arterial blood and traversing to the region of the ventricle is not necessarily an artery. The internal venous anastomosis of the superficial middle cerebral vein with the internal cerebral vein through the longitudinal caudate vein of the lateral ventricle[52], can, when filled with arterial blood, be a source of deception.

In two of the seven lesions, the arteriovenous aneurysm appears to derive is supply solely from the vessels of the choroid plexus. An ineffectual section of the choroid plexus was performed in one case (H. F., 1700/35), but the patient is alive and at work 20 years later. The other patient (E. O. N., 770/38) did not receive surgical intervention and was unchanged when last heard from 2 years after discharge.

*Basal ganglia.* On rare occasion, the artery to the basal ganglia contributes to the blood supply of the lesion. This must be recognized preoperatively, for, although these vessels are visible on the arteriogram, excision of the lesion is not surgically feasible.

*Retinal lesion.* A concomitant retinal lesion was encountered in one instance and found to conform to the association noted by WYBURN-MASON[72] of a lesion of the retina accompanied by a similar lesion of the mid-brain. It would not seem unreasonable, therefore, to further indicate the desirability of conducting an intensive search for a lesion of the deep circulation in every patient known to have a congenital angiomatous lesion of the retina.

### Case VII

Boy, age 5. Lesion of the retina and middle cerebral artery. Expectant treatment[40].

*First admission* (Nov. 16, 1938). L. E., 856/38, a 5 year old boy, who, following a head injury from a fall, developed a right hemiparesis of the face and extremities. *Examination.* In addition to the hemiparesis, an angiomatous malformation of the fundus of the left eye was noted (Fig. 31). *Encephalograms* were not remarkable. *Clinical course.* Since the significance of the retinal lesion had not been recognized, the boy was discharged on the 16th hospital day. His condition gradually improved and the hemiparesis disappeared. Two years after discharge the hemiparesis recurred abruptly, accompanied by aphasia.

*Second admission* (Sept. 12, 1946). *Examination* revealed a right hemiparesis involving facial and limb musculature, aphasia, and mental retardation. The eyeground findings comformed to the previous description 8 years before readmission. *Angiography* revealed a small arteriovenous aneurysm fed by two large branches of the middle cerebral artery (Fig. 32). The lesion was considered inoperable. When last reported 5 years following the second hospitalization, the condition of the patient was unchanged.

**Comment.** Unfortunately this lesion is not amenable to surgery.

Fig. 31. Case VII. A Wyburn-Mason lesion of the eyeground

Although the venous drainage of the arteriovenous aneurysm does not have the same surgical import as the arterial supply, a few observations might be appropriate. The arteriovenous aneurysm is preponderantly drained by the superficial venous circulation. Lesions of the anterior or middle cerebral artery drain most often into the superior longitudinal sinus, and less frequently into the middle cerebral vein, the deep venous circulation or the lateral sinus. The same is true of lesions supplied by both these arteries. On the other hand, lesions involving the posterior cerebral artery tend to discharge a proportionately greater part of their blood into the deep circulation. The

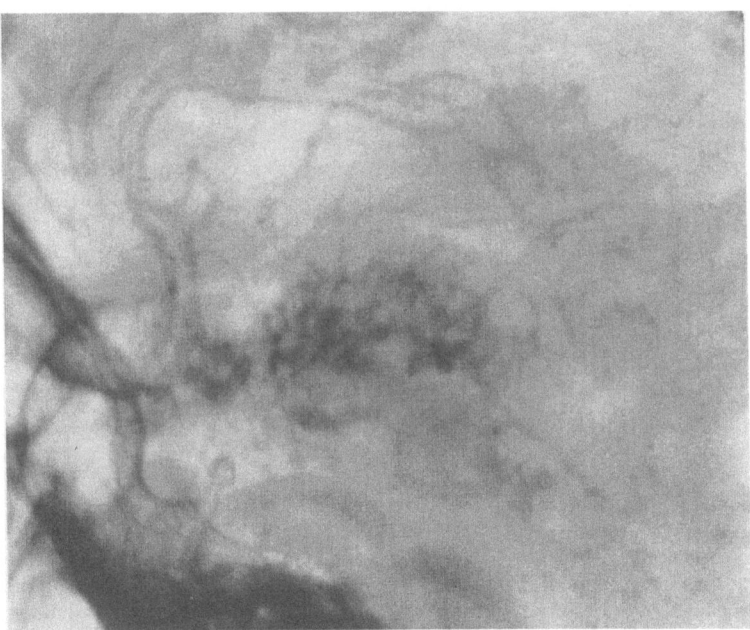

Fig. 32. Case VII. A Wyburn-Mason lesion of Middle Cerebral artery situated in the region of the sylvian fissure. Drainage to the sigmoid sinus and to the base of the skull

frequent occurrence of the superior sagittal sinus as a site of drainage necessitates the mobilization of an ample border of bone over the sinus when involvement of this channel is anticipated.

# Physical Characteristics of the Malformation

## 1. Size

Most aneurysms coming to the attention of a clinic have reached, or exceeded, the maximum size compatable with the tenuous walls of the malformation. Since the difficulties encountered at operation are often in direct proportion to the dimensions of the lesion, a small malformation, often presenting with intracranial hemorrhage, offers a relatively better surgical prognosis.

### Case VIII

Woman, age 29. Intracranial hemorrhage resulting from a lesion of the middle cerebral artery. Excision. Residual weakness of the hand [40].

*Admission* (Nov. 27, 1935). A. N., 3289/35, a 29 year old woman admitted with a history of a recent cerebral hemorrhage resulting in a right hemiplegia and aphasia. *Angiography* disclosed a small arteriovenous aneurysm in the distribution of the left middle cerebral artery (Fig 33). *Operation* (Dec. 4, 1935—twelve days after the hemorrhage). Left temporal parietal flap. After the removal of a large clot from the substance of the parietal lobe, it was seen that a significant part of the lesion had been destroyed by the bleeding. The remainder of the lesion could be excised without difficulty. *Postoperative course.* Uneventful recovery. The condition of the patient continued to improve, the aphasia disappeared, and except for a minimal residual weakness of the right hand, the patient has returned to her usual routine of living.

**Comment.** Once again, the presence of an intracerebral clot aided, rather than hindered, the excision of the aneurysm.

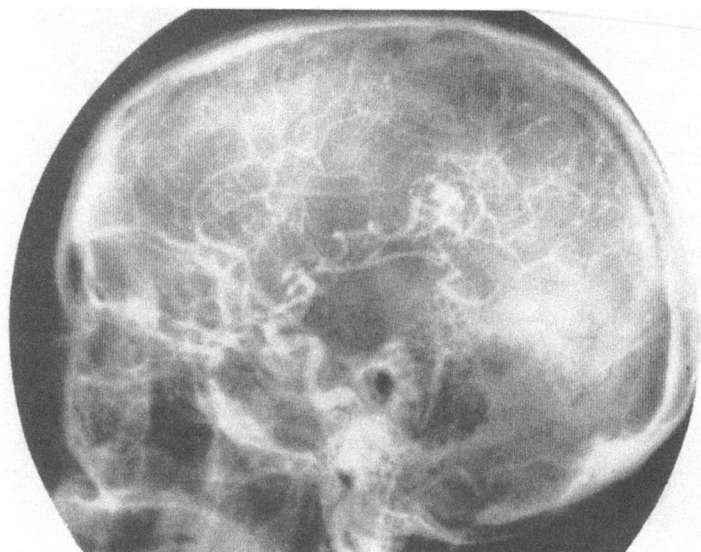

Fig. 33. Case VIII. *Left internal carotid angiography.* A small lesion in the distribution of the middle cerebral artery

Such lesions of small dimension unfortunately occur rarely. All too often the malformation is of so comprehensive a size that surgery is precluded.

### Case IX

Man, age 43. Huge lesion of the middle and anterior cerebral arteries. Exploration. Expired [40].

*Admission* (Sept. 18, 1944). G. O., 884/44, a 45 year old man admitted with a two year history of dizziness, headache and general weakness. *Examination* revealed a generalized debilitation, moderate bilateral exophtalmus, left hemiparesis with asteriognosis, bilateral papilledema with secondary atrophy, diminished visual acuity and dementia. *Skull:* pronounced increase in skull diameter, especially in the frontoparietal region. The right middle meningeal groove was enlarged and a large calcified area was present in the right frontal region (Fig. 34). *Ventriculograms:* dilatation of both lateral ventricles. The right lateral ventricle was displaced inferiorly and indented in its anterior margin by a calcified tumor (Fig. 35). *Operation* (Sept. 21, 1944). Surgery was undertaken with the preoperative diagnosis of parasagittal meningioma. At operation an enormous arteriovenous aneurysm was found in the distribution of the middle and anterior cerebral arteries, with extensive intramural calcification of the involved veins (Fig. 36). *Postoperative course.* Six weeks following discharge from the hospital, the patient died suddenly.

On the other hand, a large lesion does not necessarily abridge the life span of the patient to a greater degree than a small lesion, although the probability for increased circulatory disturbances in the tissues may be greater.

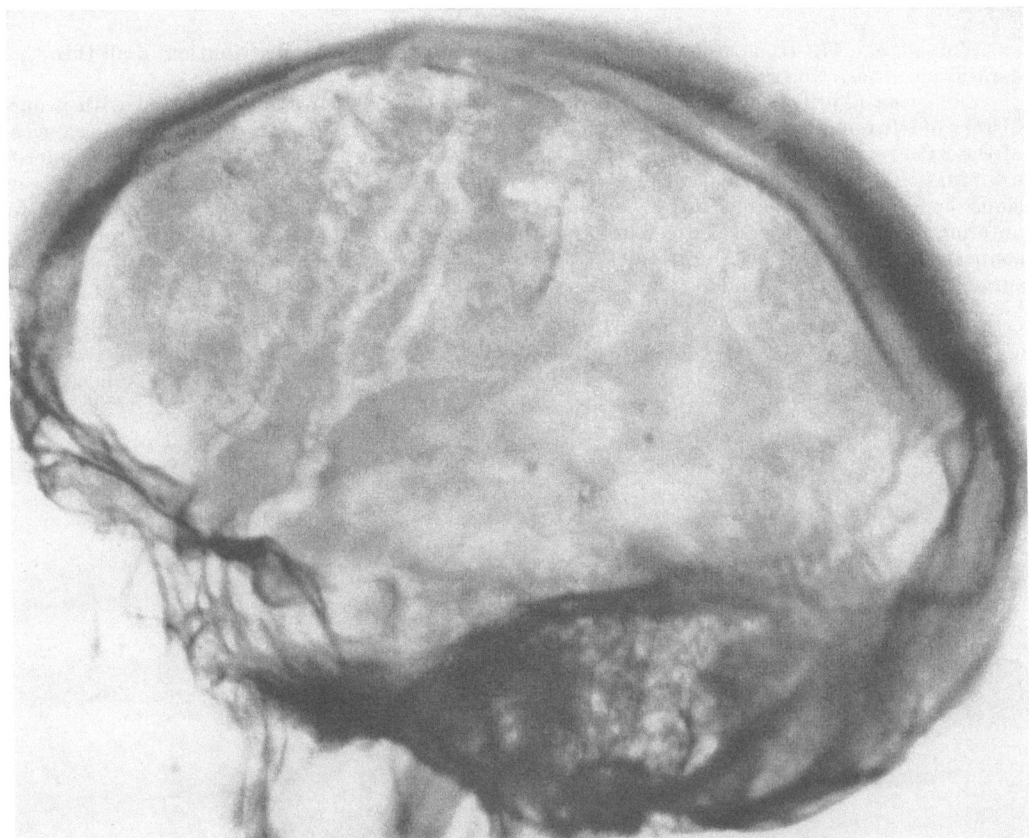

Fig. 34. Case IX. *Skull films.* The skull is thickened in the frontal and parietal regions. Middle meningeal groove is wide and tortuous. Intracranial calcifications are present in the right fronto-parietal region

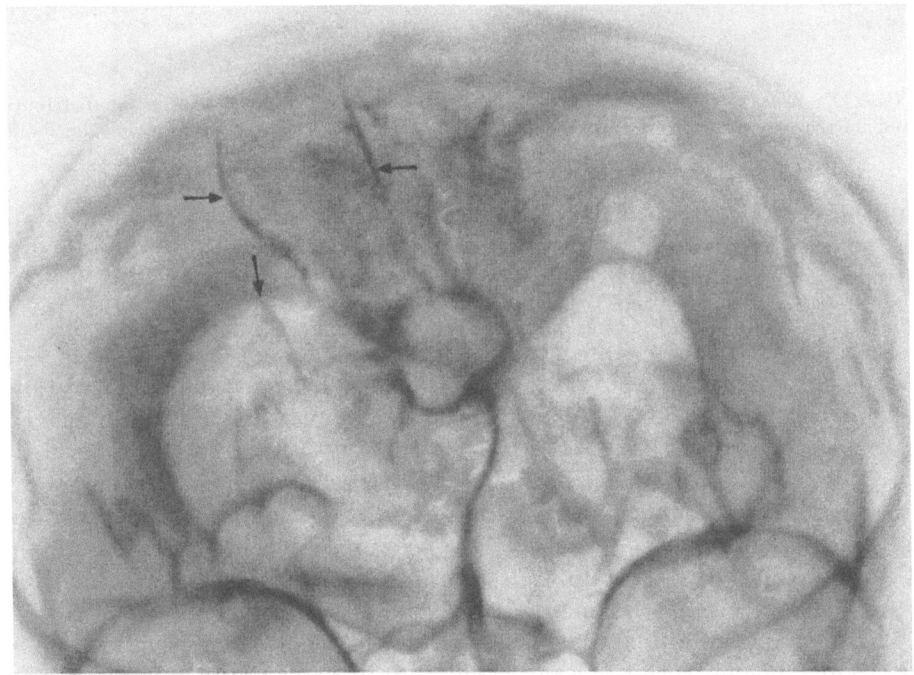

Fig. 35. Case IX. *Ventriculogram.* Moderate dilatation of both lateral ventricles. The right lateral ventricle is displaced inferiorly by the calcified tumor. (Distortion of the third ventricle was also observed)

### Case X

Woman, age 47. Huge lesion of the frontotemporal region in the distribution of all three internal carotid branches. No surgery. Survival after 20 years.

*Admission* (April 17, 1935). S. E. G., 1089/35, a 47 year old woman admitted with a one year history of left temporal headaches, blurred vision in the right eye, episodic vomiting, nausea, weakness of the left arm and leg. *Examination.* Left hemiparesis and a mild euphoria. The remainder of the examination, including visual fields, was not remarkable. *Skull.* Calvarium thickened. Sigmoid sinus bone impression especially well delineated (Fig. 37). *Angiography:* a huge lesion fed by anterior, medial and posterior cerebral arteries in the left frontotemporal region (Fig. 37). No connection with the external carotid artery. *Clinical course.* In view of the extent of the lesion, surgery was not recommended. The patient was discharged as unimproved.

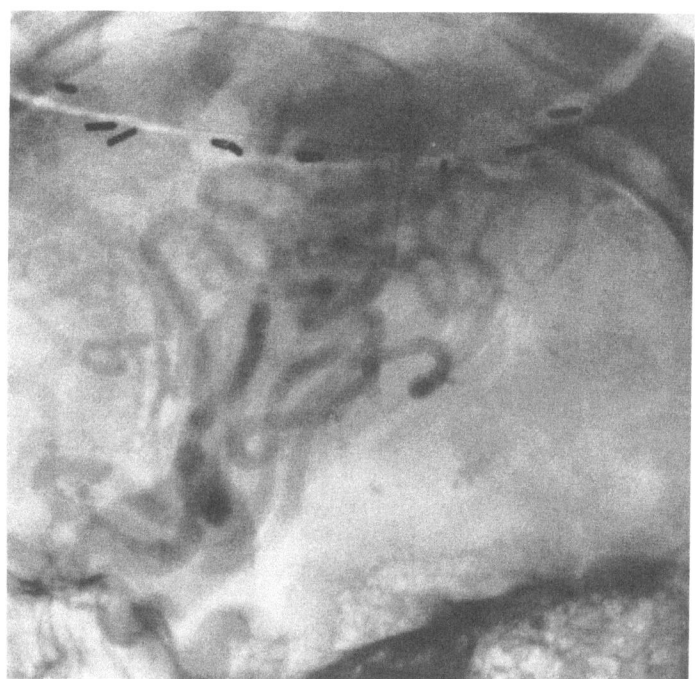

Fig. 36. Case IX. *Right internal carotid angiography* (postoperative). A huge lesion in the distribution of the middle and anterior cerebral arteries, with contrast-filling of only the arteries supplying the malformation. Extensive calcification of the walls of the vessels

**Comment.** For twenty years the clinic has received yearly letters from the patient. She was able to return to light housework and has not once had intracranial bleeding during this interval

Whether the arteriovenous aneurysm significantly enlarges with time is a question which in our experience has not yet been resolved. The anatomic composition of the angiomatous fistula, of course, remains unchanged, and while the draining veins may enlarge somewhat, it cannot be stated with certainty that the spatial relationships present at birth are materially different from that encountered at the time of surgery. A pronounced enlargement should be accompanied by correspondingly marked cardiovascular changes, which, as mentioned above, has not been observed in any of the nine patients studied with cardiac catheterization[22]. Significant alterations of size have not been observed either in a 42 year old man (K. B., 228/34) followed for 10 years (Fig. 39–41), or in a 43 year old woman followed for 20 years (Fig. 42, 43). It seems more likely that any significant tendency for expansion would be interrupted by a "blow out" of the fragile angiomatous walls. Often the so-called growth of the lesion demonstrated by angiography is artifactual, resulting from a comparison of films taken at different stages of filling (Fig. 39, 40). Incomplete filling of the early lesion may also result from a recent vascular accident or even from unaccountable spasm.

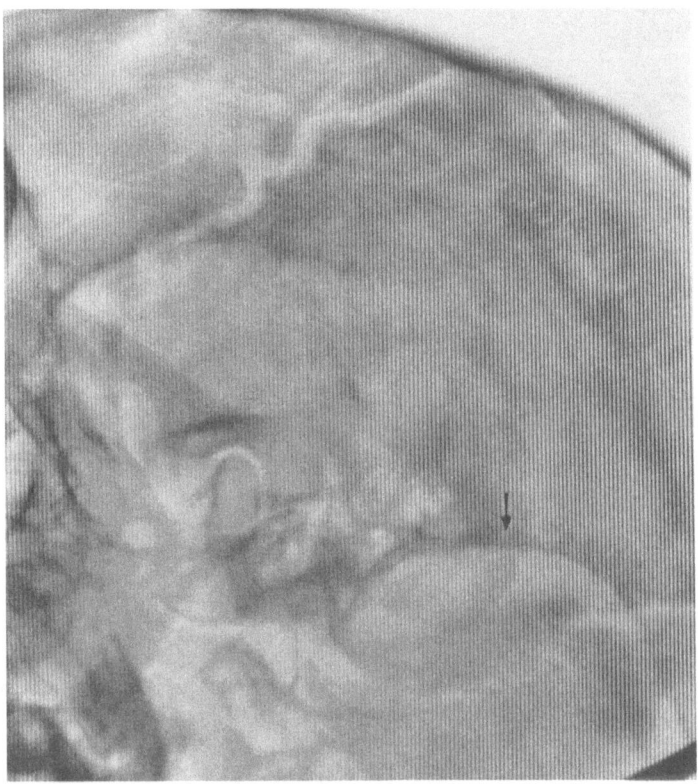

Fig. 37. Case **X**. *Skull film*. Marked meningial grooving on the right side. Note the pronounced sigmoidal sinus marking

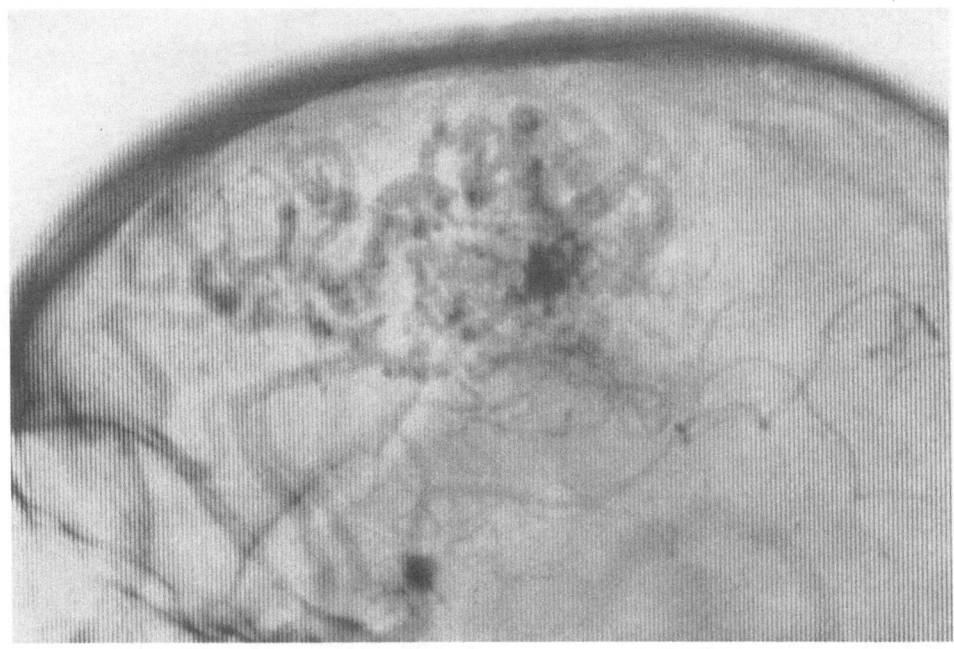

Fig. 38. Case **X**. *Left internal carotid angiography*. Both anterior cerebral arteries visualize. A large arteriovenous aneurysm situated in the left frontotemporal region. The choroid artery alone is of normal caliber. The aneurysm fills from right carotid injection as well

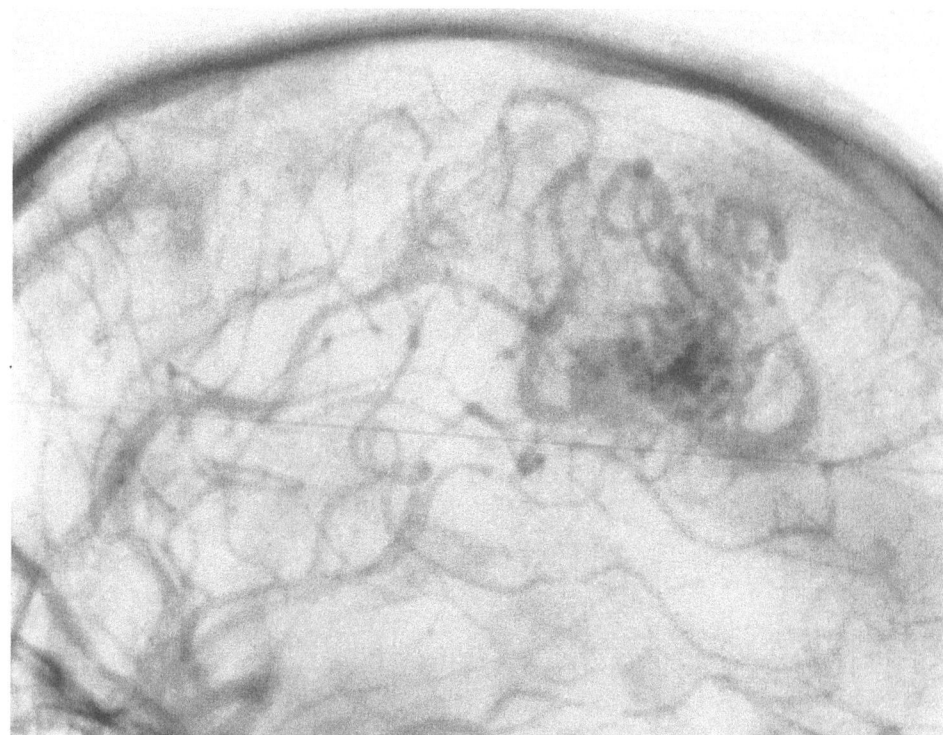

Fig. 39. K. H. B., 31 year old man, 228/34. *Right internal carotid angiography* (1934). A large parietal lesion supplied by 4 branches from the right middle cerebral artery and by 1 branch from the anterior cerebral artery. The aneurysm could be also visualized from a left internal carotid study

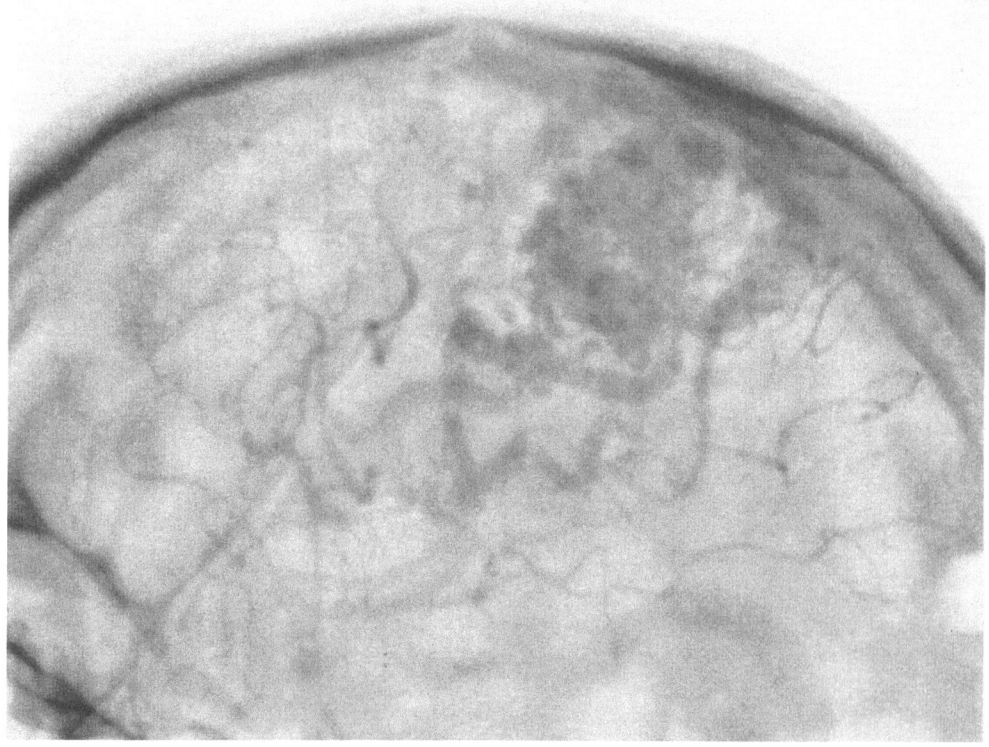

Fig. 40. Same angiogram as in Fig. 39. A later phase of filling

The problem, however, is far from settled. TÖNNIS and SCHIEFER[64] have recently demonstrated considerable increase of size in a malformation followed for 16 years by

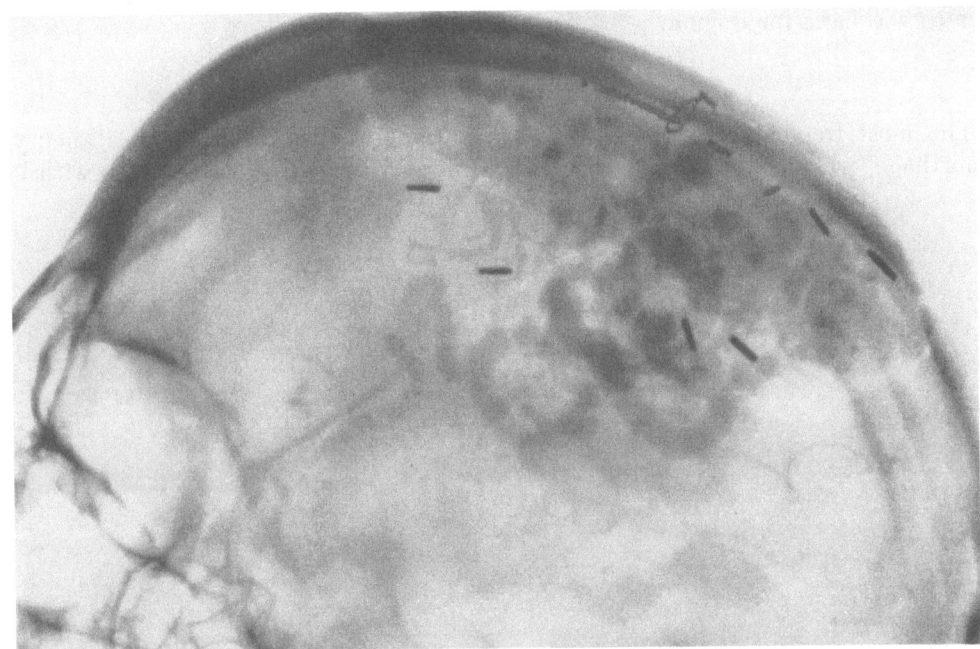

Fig. 41. K.H. B., 31 year old man, 2285/34. *Right internal carotid angiography* (1944). An exploration had been performed 10 years earlier. Apart from insignificant dilatation of the draining veins, the lesion is spatially unaltered. Note the importance in chosing for comparison the equivalent phases of filling

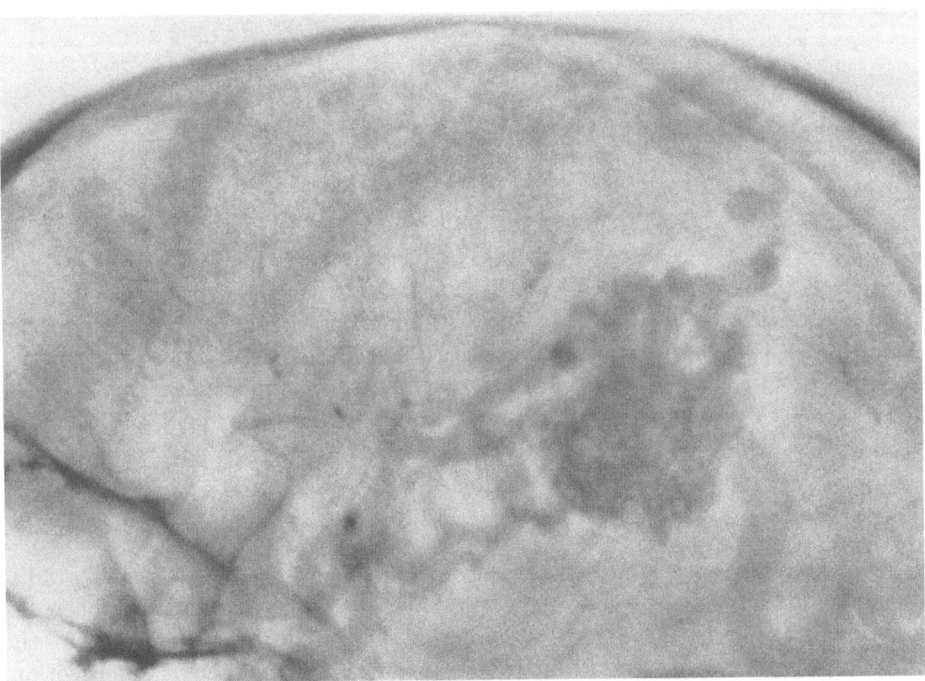

Fig. 42. J. M. A., 43 year old woman, 91/35. *Left internal carotid angiogram* (1935). A lesion in the posterior part of the left temporal region. Note the filling of only the 4 branches supplying the malformation. Venous drainage to the superior longitudinal sinus

angiographic study. Although the applicability of their case to the matter of growth is somewhat vitiated by the fact that an operative procedure (ligation of certain afferent vessels) had been performed at the time of the first admission of the patient, the case has sufficient correspondence to the problem to deter us from using our own two observations as the basis for dogma.

## 2. Hemorrhage

The most frequent and the most dreaded sequel to the arteriovenous aneurysm i hemorrhage, which in this series occurred in more than half of the patients with lesion

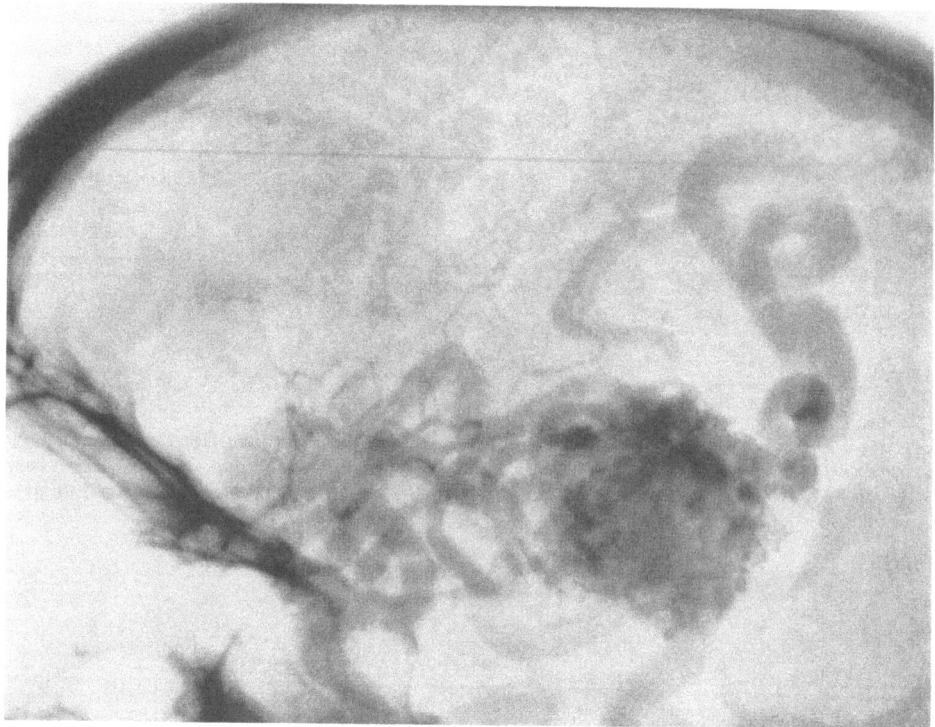

Fig. 43. J. M. A., 43 year old woman, 91/35. *Left internal carotid angiogram* (1955). No significant alteration in size after a 20 year interval

situated within the brain substance (Table 1, 2). Not infrequently repeated hemorrhage antedated definitive diagnosis. The true incidence of intracranial bleeding, moreover, must be adjusted to a higher value, since many patients succumb from their first hemorrhage before reaching a neurosurgical center, where the diagnosis can be established and recorded. Interestingly enough, about half of the bleeding occurred from lesions of small or moderate size, which confirms the observations of others [42, 63].

## 3. Thrombosis

On the infrequent occasions when thrombosis occurs, the process is gradual and usually unattended by symptoms. A recent thrombosis has never once been observed in this series at the time of surgery. Small rounded shadows often represent the cross section of large veins which have undergone degenerative changes. Calcification, however, is not an absolute indication of preexisting thrombosis, for it can be encountered both in an area of previous hemorrhage and even in abnormal vessel walls without thrombotic changes.

# Symptoms

The cardinal symptoms of the congenital arteriovenous aneurysm are, in order of frequency, epileptic attacks, intracranial hemorrhage and headaches.

**1. Presenting symptoms.** The initial symptom, as noted above, usually appears during the second or third decade. In 50 patients (40 %), the presenting symptom was some form of epileptic attack; in 48 patients (39 %), intracranial bleeding; while in the remaining 27 (20 %), the onset of symptomatology was marked by such miscellaneous complaints as headaches, migraine, dizziness, etc. (Table 1). Among the bizarre patterns was the slowly progressive hemiplegia beginning in childhood and resulting in retarded growth of the involved extremities (Fig. 44–48).

In every age group except for the third decade where intracranial bleeding predominated, epilepsy was the most prominent presenting symptom, with the male sex in the majority. The earliest presenting symptom in our material was encountered in a 14 year old boy (A. L., 640/49) who had epileptic attacks since the age of 4 (Fig. 49–52).

The initial complaint at times is episodic, punctuating symptom-free intervals lasting for many years, before the appearance of new or catastrophic symptoms brings the patient to a neurosurgical center. One of our patients (S. J. A. H., 177/39), for example, with a lesion of the left frontal lobe, had an epileptic attack at the age of 16, and after 5 symptom-free years, a subarachnoid hemorrhage. Thereafter, the patient was free from complaints for 6 years until subarachnoid hemorrhage recurred, followed by epileptic attacks of increasing frequency.

**2. Epilepsy.** Epileptiform attacks were not only the most common presenting complaint but the most frequent symptom as well (Table 2). This is in general accord with the observations of TÖNNIS and LANGE-COSACK[63]. As with the presenting symptom, epilepsy was the most usual symptom and the most common initial complaint in every decade of life except the third, when intracranial hemorrhage predominated.

From our material, it would appear that given sufficient time, almost all lesions of the cerebrum will produce some form of episodic attack. In 125 patients, epileptic attacks were present in 75 (60 %) and absent in 50. If from the latter figure we subtract 5 basal ganglion lesions, 7 lesions involving the external carotid artery in the superficial tissues, 9 lesions of the posterior fossa and 15 lesions extirpated soon after the first episode of intracranial bleeding, the total number of patients who, although candidates for epileptic attacks, did not develop the symptoms, was only 14.

The epileptic manifestations were more variable in type and frequency than those encountered in the traumatic or so-called "cryptogenic" epilepsy. The attack frequently takes the form of a jacksonian seizure. As in the experience of MACKENZIE[33] the attacks were usually focal at some time in the clinical history, but gradually spread to involve the contralateral side, accompanied by episodic loss of consciousness. Focal sensory fits were occasionally observed, as were sensory aura followed by unilateral or generalized convulsions. While not common, postictal paralysis or aphasia have been encountered. Petit mal occurred in 6 patients, invariably succeeded by generalized or focal seizures. Psychomotor attacks were encountered in 3 patients and uncinate fits in one instance.

The frequency of epileptic manifestations is most variable. Generalized convulsions are usually separated by an interval of several months or years, while pure jacksonian attacks recurred within a much shorter period and in some patients as often as several times a day.

Electroencephalography[23] may reveal no abnormal findings but more often generalized dysrhythmia will be present, usually pronounced in the affected hemisphere. If a focus be present, it is usually associated with slow waves of high voltage distributed over a comparatively large area. No pathognomonic changes in the electroencephalogram have thus far been observed.

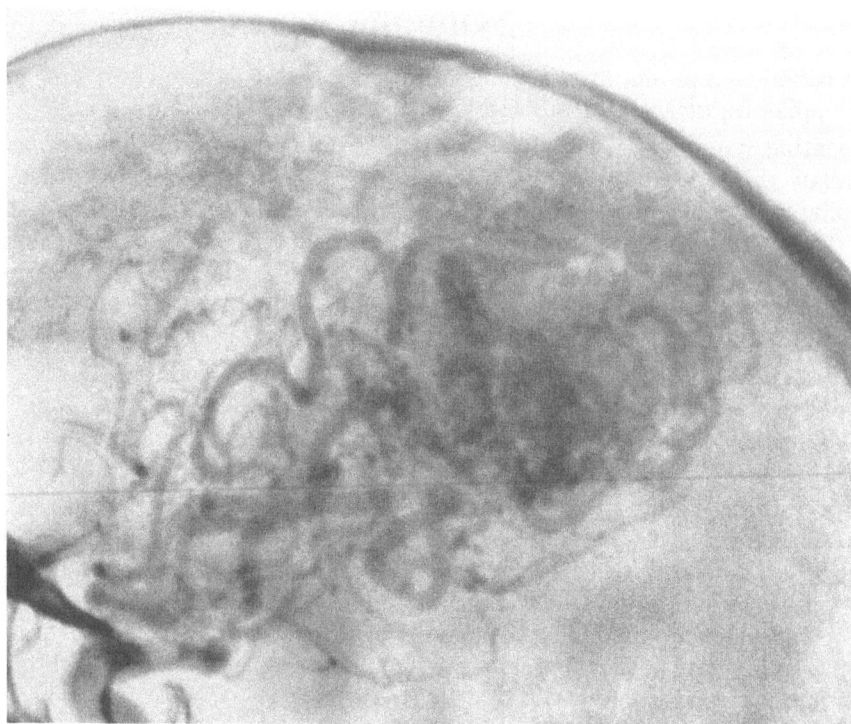

Fig. 44. M. J. S., 32 year old woman, 446/50, spastic hemiplegia since infancy. *Left internal carotid angiography* (preoperative). A lesion of the left convexity. Arterial supply is derived from 3 branches of the middle cerebral artery, which enters the aneurysm at its anterior end and from below. Contribution also from the anterior cerebral artery, which is dilated. (Drainage to the sinus sagittalis, the sphenoparietal sinus and the confluent sinus)

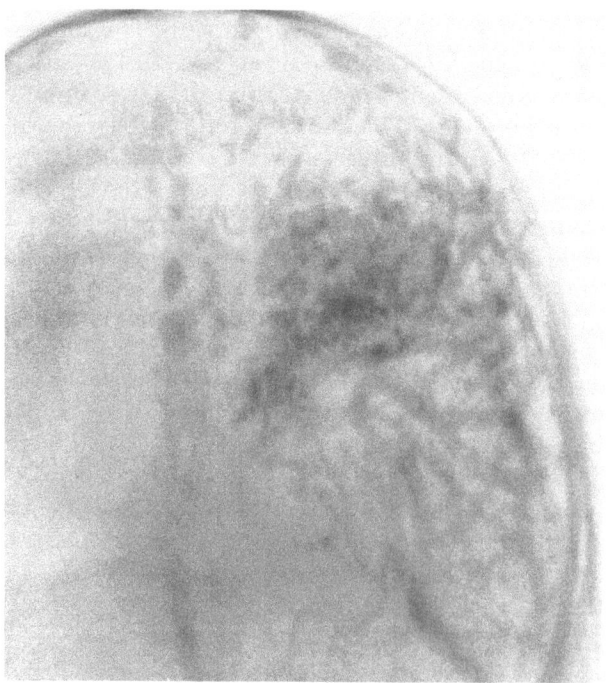

Fig. 45. See text of Fig. 44

**3. Hemorrhage.** Intracranial bleeding occurred in 63 (50%) patients, and was the most commonly encountered symptom and presenting complaint in the third decade of life when man is exposed to the greatest stress (Table 1, 2). Not infrequently, trauma preceded or precipitated the onset of the vascular accident. In 39 patients the bleeding was subarachnoidal, recurring at least once in 11 patients and in one case no less than 5 times. Subarachnoidal hemorrhage was frequently accompanied by hemiparesis or a limb palsy, usually less severe than that observed after intracerebral bleeding. Intracerebral hemorrhage, evidenced by severe motor, sensory, speech or visual field defects, occurred in 24 patients. These statistics make no allowance for the patient not undergoing surgery, who later experienced vascular insult.

Intracranial bleeding, of course, not only endangers life, but compromises cerebral function as well, especially since the lesion most often involves the middle cerebral artery.

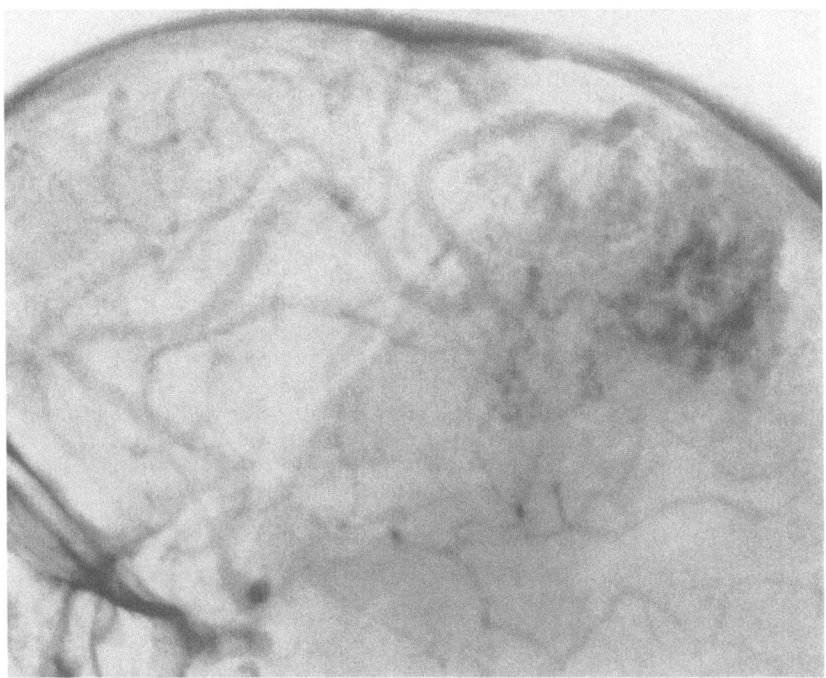

Fig. 46. M. J. S., 32 year old woman, 446/50. *Right internal carotid angiography* (preoperative). The middle and upper portion of the left-sided aneurysm visualize chiefly from the left anterior cerebral artery which fills only from right carotid injection

Unless the aneurysm is small and the clot removed without delay, serious motor, sensory or speech defects may be anticipated in proportion to the destruction of cerebral tissue.

The permanent loss of brain function ensuing after intracranial bleeding may be contrasted with the temporary loss of motor or speech function occurring after a jacksonian attack, where, unless the lesion is large, complete recovery follows the attack in a matter of minutes or hours. In the absence of clinical evidence of bleeding, slowly progressing hemiparesis, aphasia or hemianopsia invariably indicates a huge lesion capable of serious circulatory disturbances, especially local ischemia, without a concomitant elevation of intracranial pressure or gross tissue destruction (Case IX).

**4. Headache,** although frequent, is not characteristic, nor, in contrast to the experience of Tönnis and Lange-Cosack[63], did its location usually have any discernible relationship to the site of the lesion. Since the patient with an angiomatous malformation does not exhibit increased intracranial pressure without intracranial bleeding, pain cannot be attributed to the pressure phenomenon. As with most explanations the causes of headache may be quite complex, but in view of the widespread dilatation of arterial vessels,—elsewhere indicted as a cause of headache,—one need not search far for at least a convenient explanation.

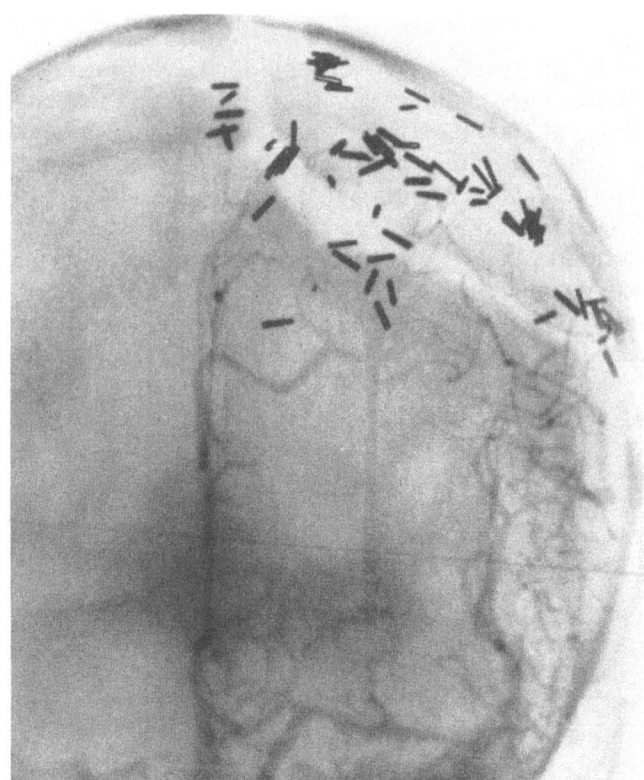

Fig. 47. M. J. S., 32 year old woman, 446/50. *Left internal carotid angiography* (postoperative). No trace of the aneurysm remains. Both anterior and posterior cerebral arteries now fill. The middle cerebral artery has returned to normal caliber. Two years after operation the patient was able to walk. The postoperative aphasia has almost completely disappeared

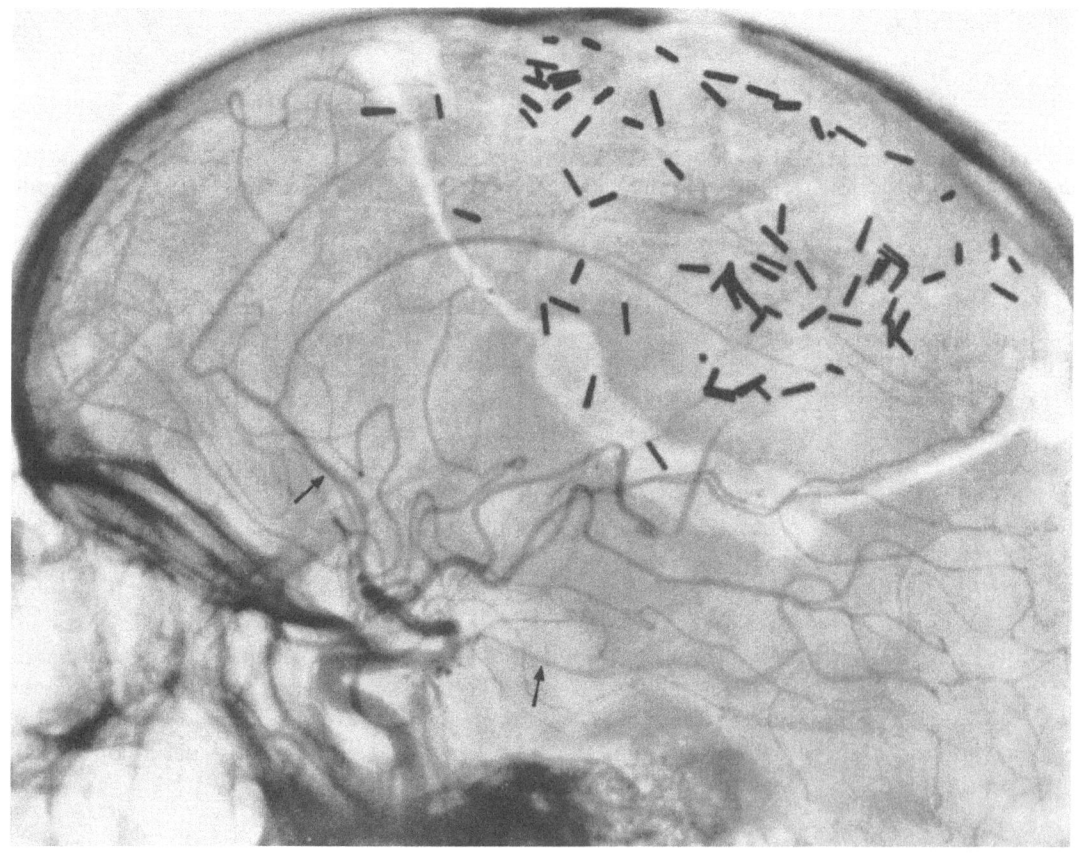

Fig. 48. See text of Fig. 47

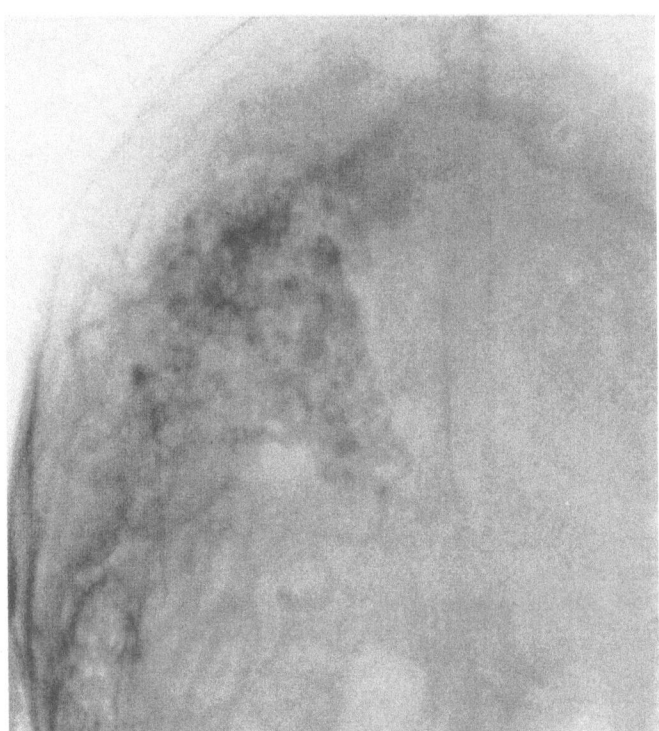

Fig. 49. A. H. L., 14 year old boy, 640/49. Epilepsy since the age of 4. *Right internal carotid angiography.* Lesion of the convexity, fed by the middle cerebral artery. Drainage to the superior sagittal sinus. The anterior cerebral artery does not visualize

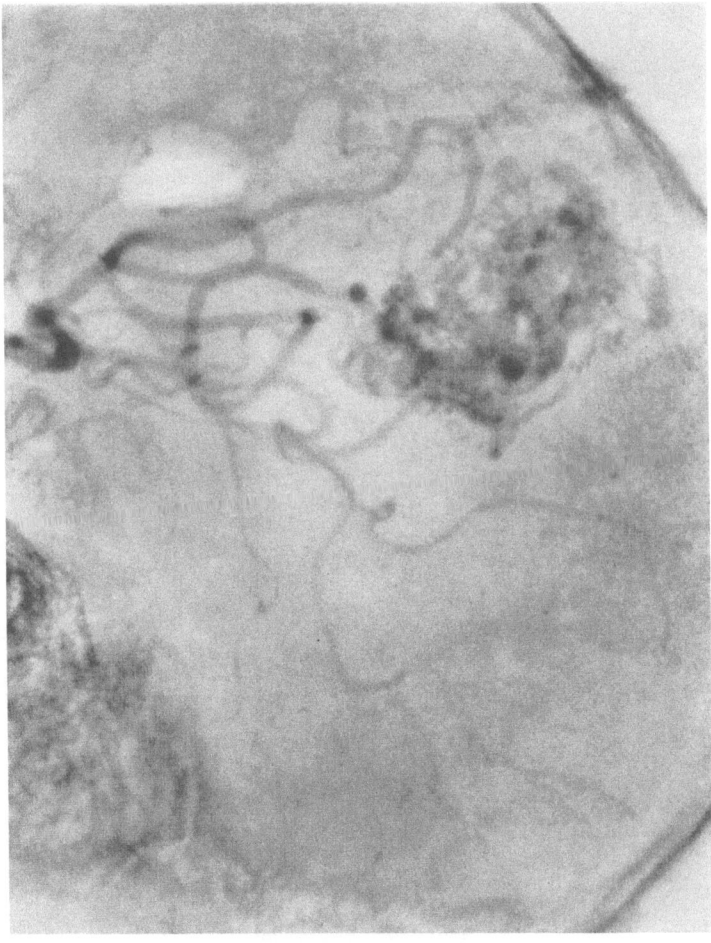

Fig. 50. See text of Fig. 49

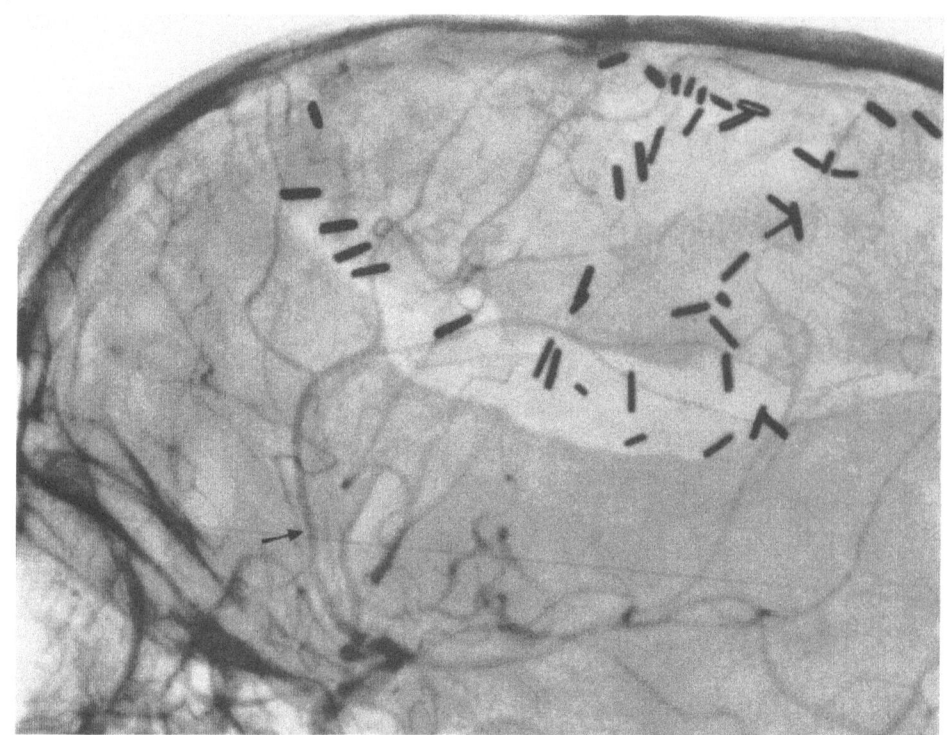

Fig. 51. A. H. L., 14 year old boy, 640/49. *Right internal carotid angiogram* (postoperative). The aneurysm has been removed. The dilated vessels have returned to normal caliber and the anterior cerebral artery now visualizes. Minimal weakness in the left extremities 6 years after surgery. Occasional epileptic attack, but the frequency is greatly diminished. Good result

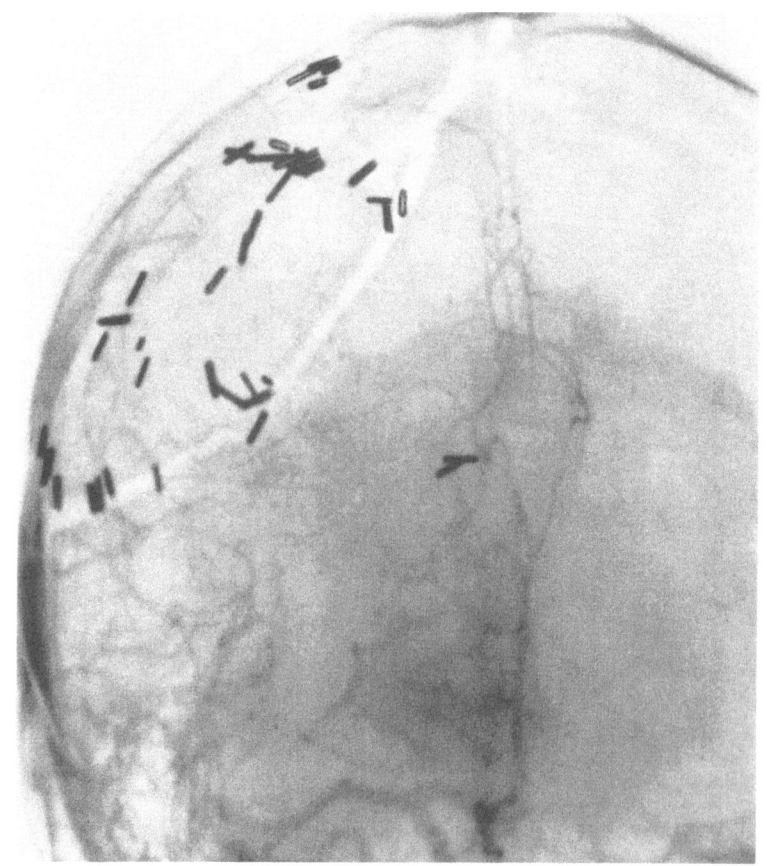

Fig. 52. See text of Fig. 51

Migraine has been noted in 4 patients, two of whom were treated for many years for this complaint before the appearance of additional symptoms directed attention to the etiology. A certain prominence has been attached in the literature[2, 12, 24] to the association of migraine with lesions of the occipital lobe. Among our four cases, two lesions were located occipitally.

**5. Exophtalmos,** although rare, may be of some diagnostic value when unassociated with increases in intracranial pressure. If unilateral, the orbital protrusion will always be situated ipsilateral to the lesion, while if bilateral, the exophtalmos will be more pronounced on the side of the lesion. A mild bilateral exophtalmos was observed in one patient, and in another the exophtalmos was ipsilateral to an occipito-parietal malformation. Pulsating exophtalmos has been reported in the series of TÖNNIS and LANGE-COSACK[63].

**6. Mental changes.** Psychic disturbances have long been recognized as a concomitant to the arteriovenous lesion, particularly in cases of long standing. In this series mild mental changes were noted in one-fourth of the patients and pronounced changes in another one-fourth, testifying to the impact of the aneurysm on cellular function. Moreover, dementia requiring institutionalization was encountered.

**7. Subjective murmur.** All of the lesions involving the external carotid artery were associated with subjective auditory disturbances except in one instance where the duration of the lesion had been brief. On the other hand, only 4 patients with malformations of the internal carotid circulation were troubled with these noises.

# Signs

**1. Murmur.** Although the occurrence of a murmur in external carotid malformations had been recognized since early in the 19th century, STEINHEIL[59] was the first to describe the association of bruit with a lesion of the brain substance. In the following years great importance was attached to this sign. CUSHING and BAILEY[7], reporting the presence of a murmur in 8 of their 9 patients, stressed its frequency and diagnostic import. Later studies[8, 51, 58] indicated that the murmur is not as common as previously supposed.

In this series a bruit was heard in 21 patients. Furthermore, it was associated with 11 of the 12 aneurysms of the external carotid artery, while in only 10 lesions of the internal carotid artery could this sign be elicited. Indeed, in the latter occurrence, the bruit invariably indicated a huge aneurysm so that the sign was almost of pathognomonic significance, although it must be remembered that a bruit may be encountered in vascular meningiomas, saccular aneurysms and even in the normal child.

The murmur is present during a great part of the cardiac cycle (Fig. 53, 54), but its accentuation during systole may give the impression that the bruit is solely systolic in occurrence. When the arteriovenous aneurysm is situated in the superficial tissues, the murmur is best auscultated over the angiomatous tissue and is propagated along the course of the draining vessels, diminishing in intensity in proportion to the distance from the malformation. The murmur *(bruit cataire, schwirrendes Geräusch)* has been compared with the following sounds: a fly in a paper box (BURGESS), mill race (RIBES), water rushing over a dam (SEEGER I), whistling of the wind (SEEGER II), bellows (WILLAUME), forcing air through a syringe (BAYER), water streaming through a metal tube (LARREY)[5].

**2. Thrill** *(fremissement vibratoire)* is a concomitant to most of the murmurs of the external carotid artery, while it is absent in the deeper lesions. The thrill is perceptible during systole directly over the communication and diminishes as the distance from the fistula increases.

**3. Papilledema** was stated in the earlier literature to be a frequent sign. The experience of this clinic cannot corroborate this assertion, for a choked disc, like other signs of increased pressure, was usually absent unless preceded by intracranial hemorrhage, when other fundus changes occurred as well. Large lesions of the posterior fossa may produce pressure symptoms, and in 2 of the 9 posterior fossa malformations a severe papilledema

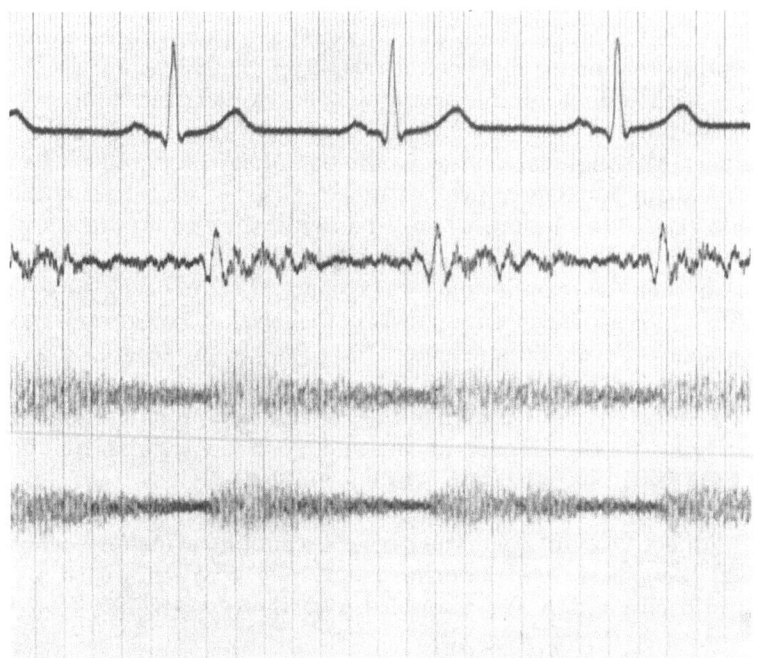

Fig. 53. G. P. P., 612/47, lesion of the external carotid artery. The murmur registers before the middle of systole and continues to the end of diastole (Dr. H. DAHLSTRÖM)

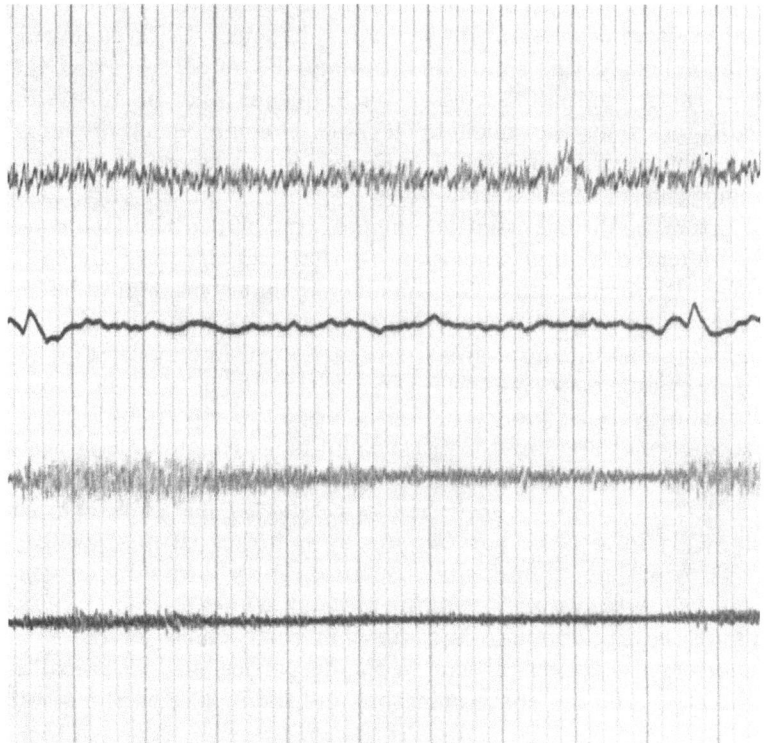

Fig. 54. J. M. A., 91/35, lesion of the internal carotid artery. The murmur persists from the last half of systole to the beginning of diastole (Dr. G. ÖHRBERG)

was present, resulting in blindness in one instance. The remaining 7 patients exhibited moderate papilledema of a few diopters.

**4. Posterior fossa signs.** Pathognomonic signs were not encountered in the 9 patients with lesions of the posterior fossa. When, as in 7 of the patients, the malformation was present in the cerebellum, fed most often by the posterior cerebral and posterior cerebellar arteries and less frequently by either the anterior cerebellar or the posterior choroid artery, general pressure symptoms were often in evidence, together with cerebellar disturbances of gait and posture. In the remaining 2 cases the lesion was situated in the cerebellopontine angle, resulting in irritation of the adjacent cranial nerves. Trigeminal neuralgia and vertigo constituted the principal pattern in one patient, and facial tic with tinnitus was encountered in the other.

# Radiography

**1. Skull.** In the presence of an arteriovenous aneurysm involving the external carotid artery, the skull films may show increased vascular markings, especially pronounced

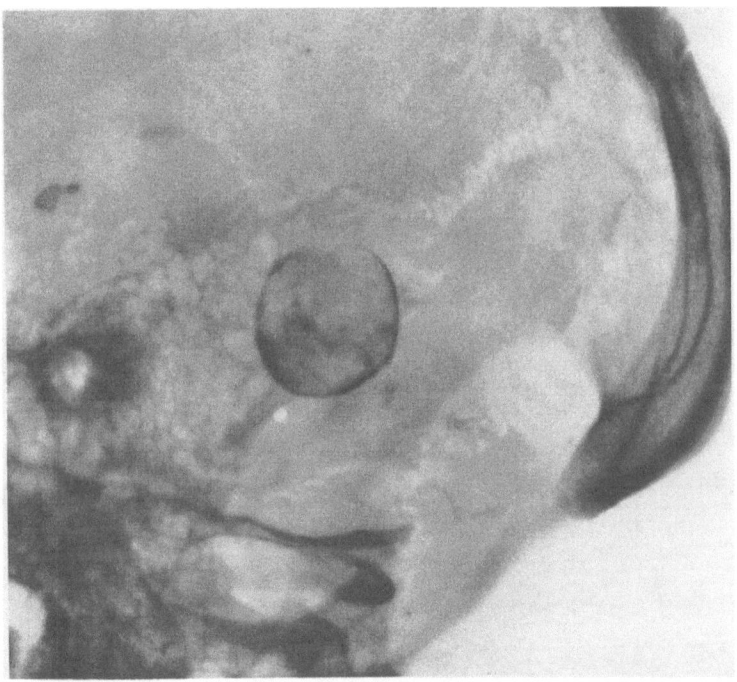

Fig. 55. J. R. V. W., 35 year old man, 1925/30. *Skull films.* A calcified mass situated in the posterior fossa on the left side, probably lying in the tentorial region, medial to the squama temporalis. The malformation was removed in 1932. The patient is alive in 1955 and able to work

in the middle meningeal groove; enlargement of the foramen spinosum; increased vascularity of the bone; enlargement of the longitudinal sinus; and in cases of long standing, a diffuse thickening of the calvarium (Fig. 34).

Lesions of the internal carotid or vertebral arteries are not usually associated with bone changes, despite the vascularity of the bone evident at surgery. Calcification, chiefly occurring in sites of previous hemorrhage (Fig. 55) is more common than either intramural calcification (Fig. 34) or calcification of an intraluminal thrombosis. The large oval intracerebral calcifications cited by Tönnis and Lange-Cosack[63] are pathognomonic for this lesion (Fig. 55). Other variants of calcification have also been observed (Fig. 56–60).

**2. Encephalography** frequently demonstrates atrophy, either generalized or confined to the side of the lesion. Large aneurysms may produce some shift of the ventricular system towards the normal side, but the displacement is slight and invariably associated

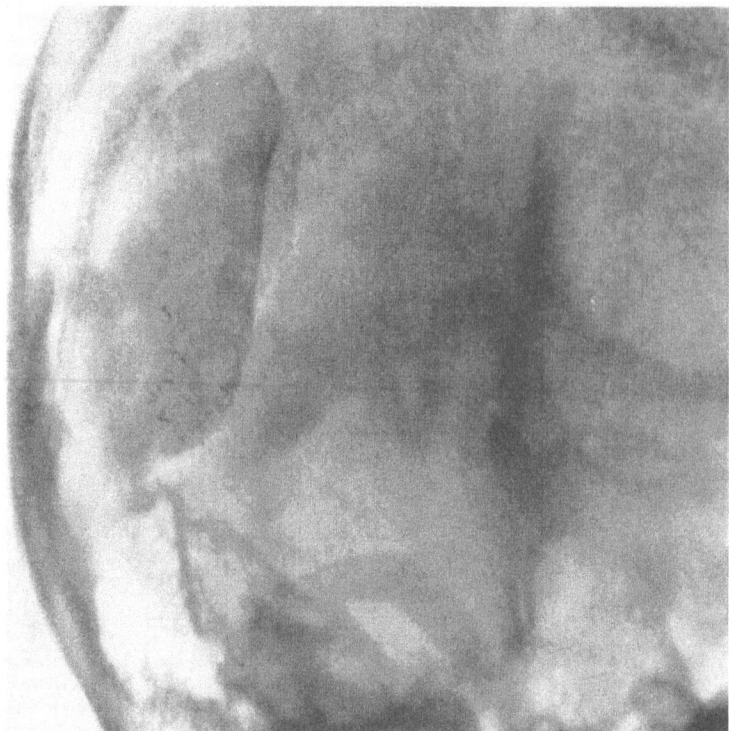

Fig. 56. E. M., 35 year old woman, 229/49. *Left internal carotid angiogram* (preoperative). Unusual form of arteriovenous aneurysm consisting of a large sac with enormous dilatation of the middle cerebral artery and almost complete absence of filling of other cerebral arteries. Epilepsy since the age of 10

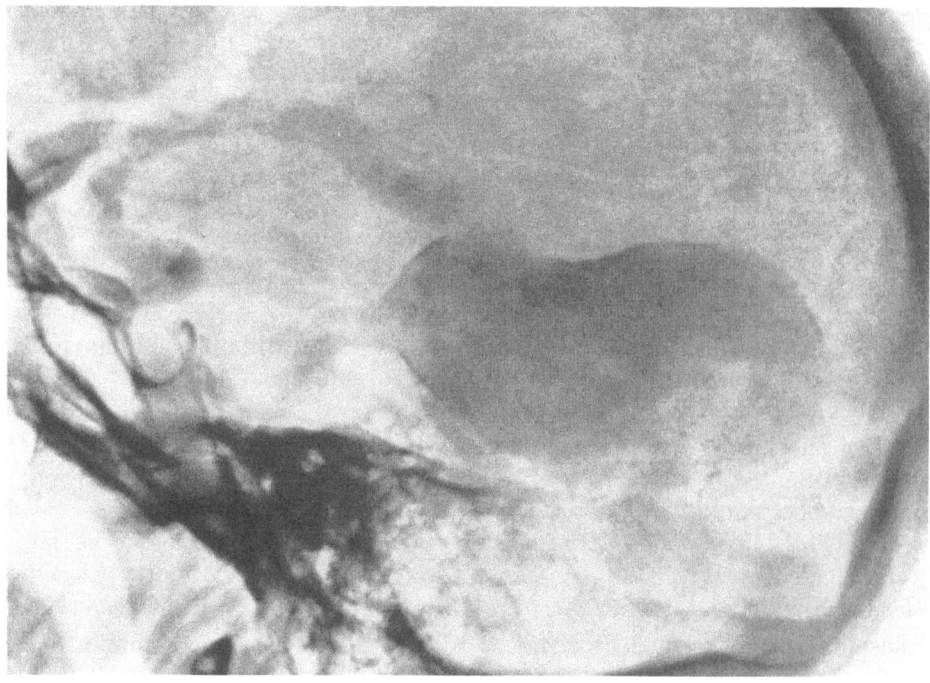

Fig. 57. See text of figure 56

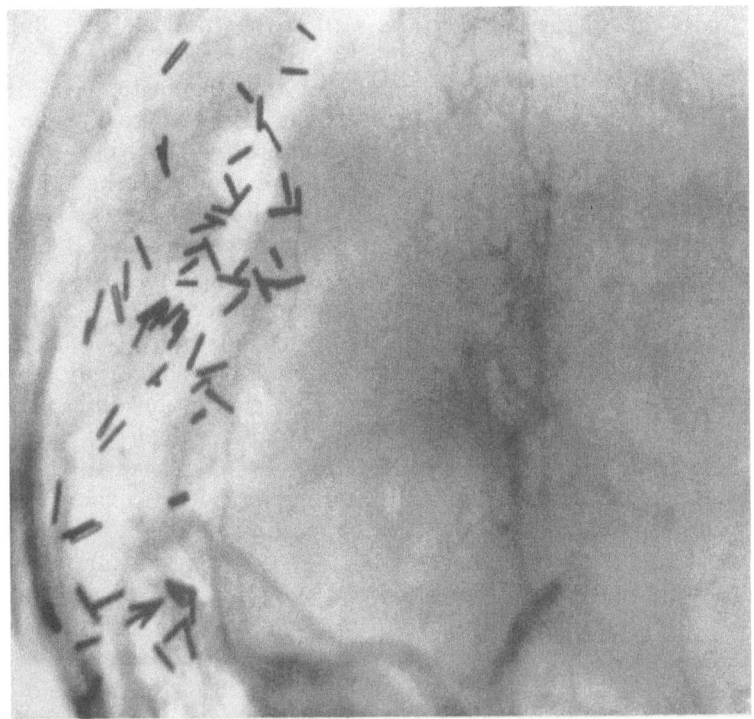

Fig. 58. E. M., 35 year old woman, 229/49. *Left internal carotid angiogram* (postoperative). The malformation has been removed. According to the last report the patient still has an occasional epileptic fit. Otherwise she feels well

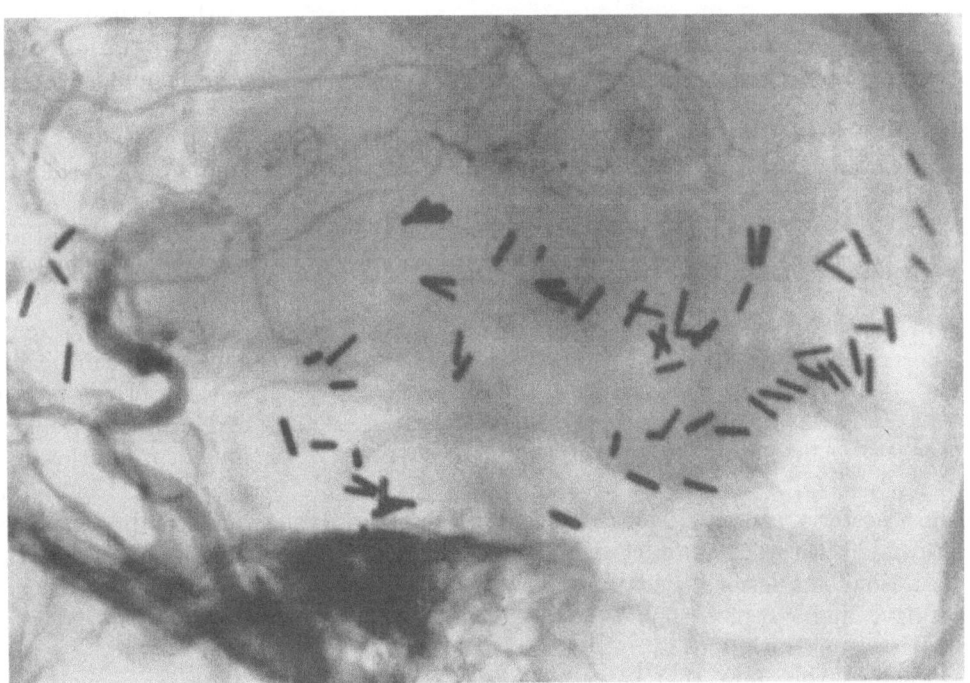

Fig. 59. See text of figure 58

with evidence of atrophy on the side of the lesion. A porencephaloid cavity may be encountered following an old intracerebral hemorrhage. The encephalogram will sometimes reveal an extension of the lesion to the wall of the ventricle (Fig. 35).

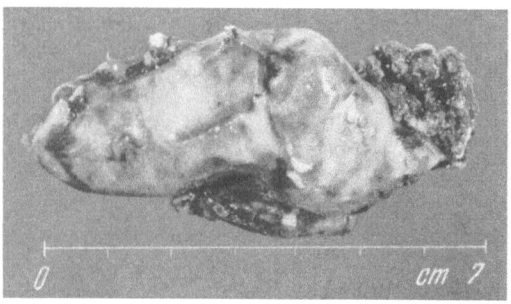

Fig. 60. E. M., 35 year old woman, 229/49. Operative specimen

**3. Arteriography** is by far the most important of the diagnostic procedures. At the Serafimer Hospital great progress in this field was made after the examination was taken out of the realm of the neurosurgeon and placed in the hands of the radiologist.

"Triurol" is employed in this clinic as the contrast medium, in accordance with the technique described by LINDGREN[32]. Percutaneous injection is used both for carotid and vertebral angiography. Because of the increased circulation time, exposures must be rapid. Errors can be encountered when sole reliance is placed in the lateral study, for soon after injection the dye may quickly find its way over to the contralateral side, so that the appearance of the aneurysm on the opposite side might mislead the examiner in determining the side of the lesion. This mistake cost one of our patients (K. J. L., 329/39) an unnecessary operation (Fig. 61–63). For this reason, apart from their value in other respects, frontal views serve as a double check and, indeed, should always be included in the examination of any patient selected for surgery.

To undertake surgery without a full examination is to hazard the life of the patient. An ipsilateral external carotid study should always be performed in a cerebral lesion visualized by the common carotid study and a positive external carotid examination is always an indication for a survey of the contralateral side. Difficulties encountered in localizing a lesion of the posterior cerebral artery may require a repetition of the procedure, but fortu-

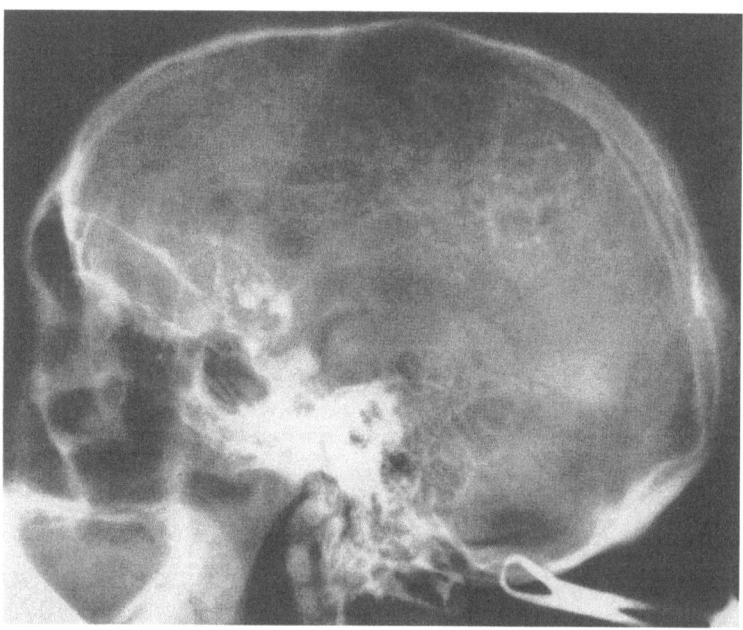

Fig. 61. K. L., 45 year old man, 329/39. *Left internal carotid angiogram.* The right-sided lesion visualized by the left internal carotid study, was thought to situated on the left. The error was discovered at surgery

nately most posterior cerebral lesions have contributions from the middle cerebral artery and will be revealed in the common carotid examination. A vertebral angiogram is indicated in all posterior fossa lesions visualized by common carotid angiography and *vice versa*. Postoperative angiography should be performed routinely to evaluate the success of the operative procedure.

It is evident, therefore, that before a complete pre- and postoperative survey can be assembled concerning the size of the lesion, its location, the participating vessels, and the

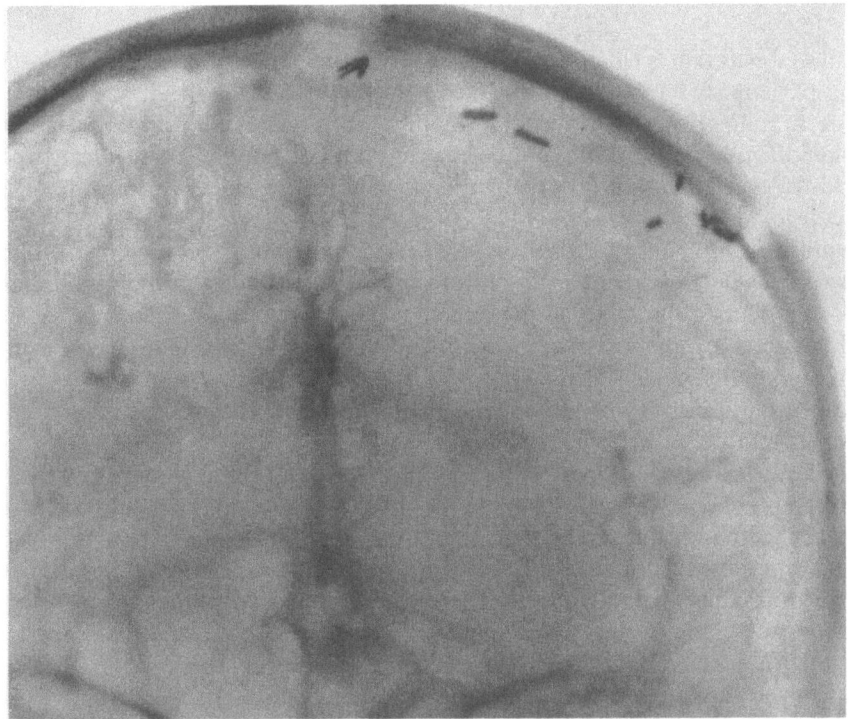

Fig. 62. K. L., 45 year old man, 329/39. *Right internal carotid angiogram*. The parietal malformation is fed by the pericallosal and cingular arteries. Drainage chiefly to the longitudinal sinus

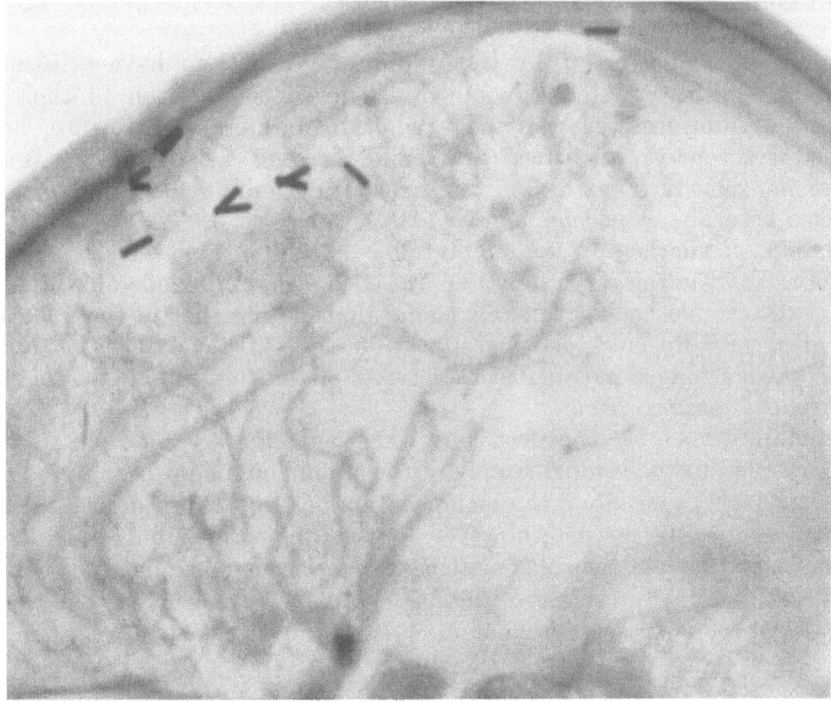

Fig. 63. See text of Fig. 62

efficacy of surgery, a great many angiographic studies must be performed. The tremendous advantages in entrusting these procedures to the radiologist can be readily appreciated.

# Differential Diagnosis

**1. Saccular aneurysm.** Until angiographic evidence to the contrary is forthcoming, every young patient with intracranial bleeding must be presumed to have either a saccular aneurysm or a congenital arteriovenous aneurysm. The earlier the bleeding, the more probable the diagnosis of arteriovenous aneurysm. Although sometimes appearing as a sequel to intracranial bleeding, epilepsy is unusual in saccular aneurysm. Bruit, likewise, is exceptional.

**2. "Cryptogenic" epilepsy.** The bizarre variations of "cryptogenic" epilepsy can coincide at times with the irregular symptom patterns observed in arteriovenous malformations. Long remissions of the generalized convulsions, alternations of focal and generalized attacks, or inconsistency of aura should suggest to the physician the possibility of an angiomatous malformation.

**3. Astrocytoma.** A slowly growing fibrillary astrocytoma producing jacksonian seizures may simulate an arteriovenous aneurysm.

**4. Posterior fossa tumor.** Without an angiographic examination it is often difficult to differentiate preoperatively between a vascular and a neoplastic posterior fossa lesion. In recent years we have come to rely on the vertebral angiogram to provide useful information concerning posterior fossa neoplasms and expect, therefore, to increase the accuracy of the preoperative diagnosis of all posterior fossa lesions.

**5. Arteriovenous aneurysm.** In a young patient the history of epileptic seizures, particularly of the jacksonian type, associated with intracranial bleeding with or without a bruit, makes the diagnosis of an arteriovenous aneurysm almost a statistical certainty.

All matters of doubt will be promptly resolved by a complete angiographic examination!

# Treatment

## A. Early Procedures

**1. Decompression** was noted by CUSHING and BAILEY[7] to have "distinctly beneficial effects". DANDY[8] limited the indications for decompression to those instances when pressure symptoms were present. Increased intracranial pressure, however, is unusual in this lesion, except after intracranial bleeding, where the merits of decompression are notoriously suspect. For this reason, decompression of the arteriovenous aneurysm has been abandoned for many years.

**2. Radiation.** Radiotherapy was employed by CUSHING[7] but he saw little value in the procedure. Early in this series radiation was tried on several inoperable lesions without the slightest discernible benefit. In this connection, noting that in the 1955 Hunterian lecture Mr. POTTER[43] lingered too suggestively on the statistical merits of radiotherapy, we wish to take a strong stand against the revival of this ineffectual form of therapy for the arteriovenous malformations.

**3. Afferent ligation.** On the occasions when angiography indicates that a single artery has provided to the arterial supply of the lesion, one might be tempted to attempt to ligate that artery, especially if the malformation lies in a surgically inaccessible area. This temptation must be resisted, however, for the angiogram is not infallible, and we have at times noted the presence of contributing vessels which had not been demonstrated by the contrast study. Furthermore, there is no absolute guarantee that the vessel to be ligated at surgery is the vessel which offends.

### Case XI

Woman, age 20. Lesion of the right temporal lobe, ostensibly fed by the right posterior cerebral artery. Ligation of the right posterior cerebral artery. Ineffectual result.

*Admission* (Oct. 5, 1953). E. I. H., 752/53, a 20 year old woman with a history of two episodes of subarachnoidal bleeding within the halfyear preceding admission. *Examination* disclosed a slight left facial paresis. The remainder of the neurological examination was not remarkable. *Angiography* of the right internal and external carotid, the right vertebral and the left internal carotid arteries: an arteriovenous aneurysm, probably situated in the medial aspect of the temporal lobe, fed by the right posterior cerebral artery (Fig. 64–66). *Operation* (Nov. 12, 1953). A large parietotemporal bone flap. Ligation of the posterior communicating artery and the dilated posterior cerebral artery. *Clinical course.* Following surgery the patient developed a right occulomotor paresis which diminished by the time of discharge. Her condition was otherwise unchanged. *Postoperative angiography.* No material change in the size of the lesion. The malformation is now supplied by small choroidal branches and by branches from the basilar artery (Fig. 67–70).

**Comment.** The angiogram does not always provide complete information concerning the afferent blood supply and can be especially misleading in lesions involving either the

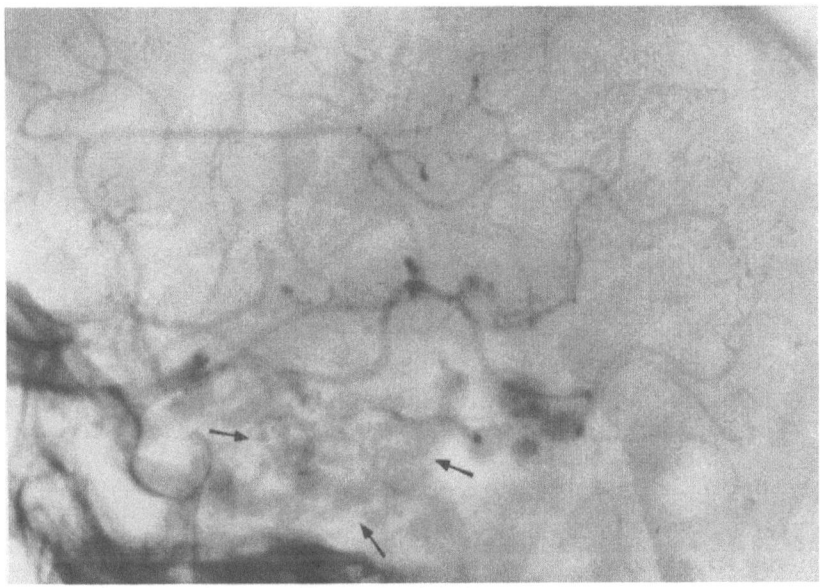

Fig. 64. Case XI. *Right internal carotid arteriogram* (preoperative). A walnut-sized arteriovenous aneurysm is situated behind the dorsum sellae, fed by the dilated posterior cerebral artery. The other carotid branches do not visualize. The malformation is drained by the basilar vein

choroidal, vertebral or posterior cerebral arteries. Ligation of the afferent vessels, therefore, is a procedure ill-advised.

A unique situation was encountered in one lesion (N. R., 558/36) involving a large branch of the anterior cerebral artery, where a change of color could be demonstrated following compression of the afferent artery. The vessel was ligated and the patient has had no further trouble. This fortuitous occurrance has not since been duplicated.

**4. Ligation of the surface vessels.** Obliteration of superficial vessels of the malformation, even more dubious in rationale than that of the above procedure, was tried in 5 patients without the remotest evidence of success. Because of its disregard for certain obvious anatomical facts, the method is categorically condemned.

**5. Carotid ligation.** Ligation of the carotid artery has been extensively employed in the past at many neurosurgical clinics, the largest series (14 cases) reported by Tönnis and Lange-Cosack[63]. Elsewhere, the paucity of published results suggests the worthlessness of the procedure.

In effect, ligation of the common carotid has an immediate effect on $a'$, whose blood supply is now derived from collateral flow through the external carotid system and the communicating system of the circle of Willis. Three conditions may be postulated, with dependence on the extend to which the collateral communications can dilate (Fig. 4, $k$ is assumed to be situated in the proximity of $f$).

Type I. If the communicating system cannot dilate sufficiently to restore $a'$ to its pre-ligation size, then not only will the flow of blood to $f$ be diminished, but the area of the brain supplied by $b'$ and $k$, formerly only scantily supplied, will now receive almost nothing. The blood flow through the fistula will be diminished, to be sure, but the result

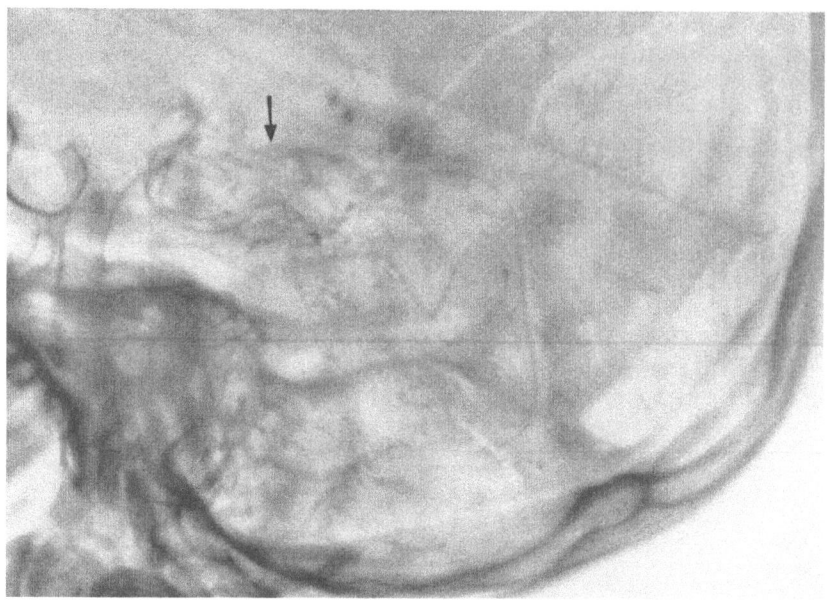

Fig. 65. Case XI. *Right vertebral angiogram* (preoperative). The lesion is fed by the posterior cerebral artery No other supply to the malformation can be demonstrated

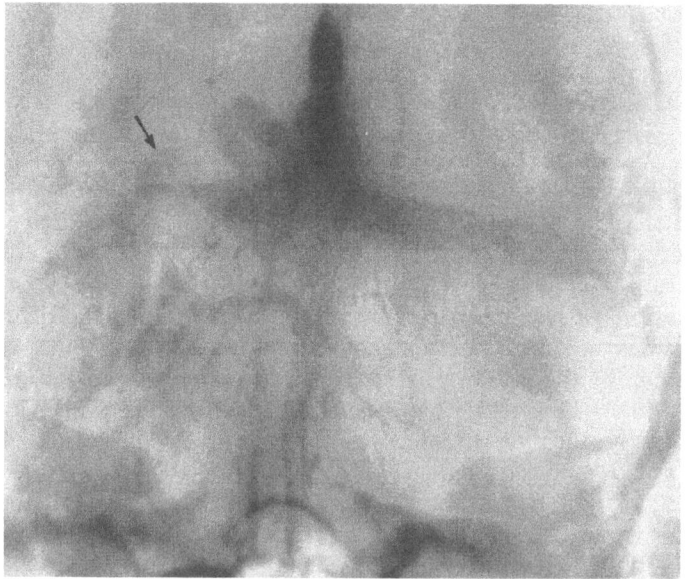

Fig. 66. See text of Fig. 65

will be virtually the same as ligating all the branches of the $a'$ and $k$ systems, or, in other words, a physiological hemispherectomy.

Type II. If the communications can dilate to the extent that the blood supplied to $a'$ will be unchanged, then the relationship of the system is unaltered and nothing will have been accomplished.

Type III. If the communications dilate slowly, the resulting pattern will resemble type I until the communicating system has restored $a'$ to its former size; thereafter the system will resemble type II.

In 6 patients ligation of the common carotid was performed.  Hemiplegia ensued in 4 patients (type I), requiring the removal of the ligature and in half the patients the

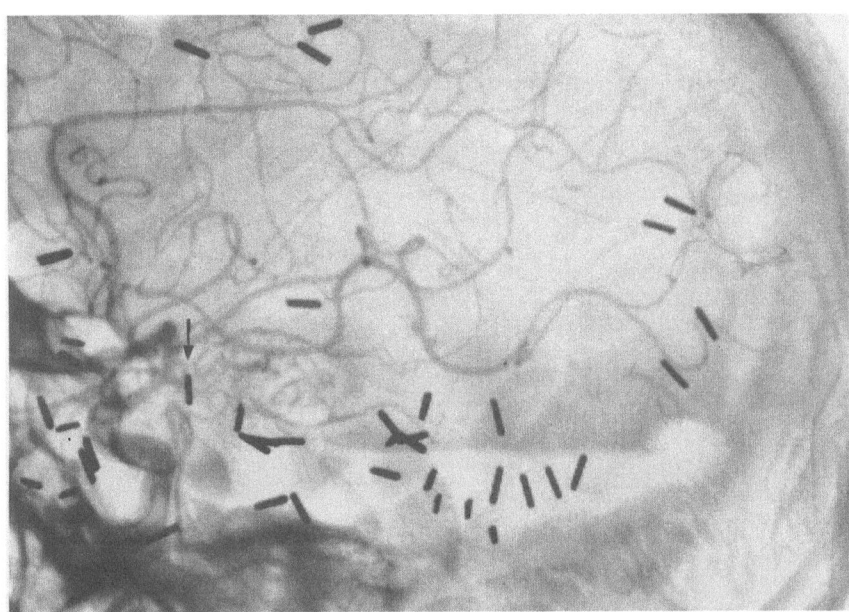

Fig. 67. Case XI. *Right internal carotid angiogram* (postoperative).  At surgery the right posterior cerebral and the posterior communicating arteries had been ligated.  The malformation does not visualize in its anterior part, but note the the filling from the small branches of the choroidal artery

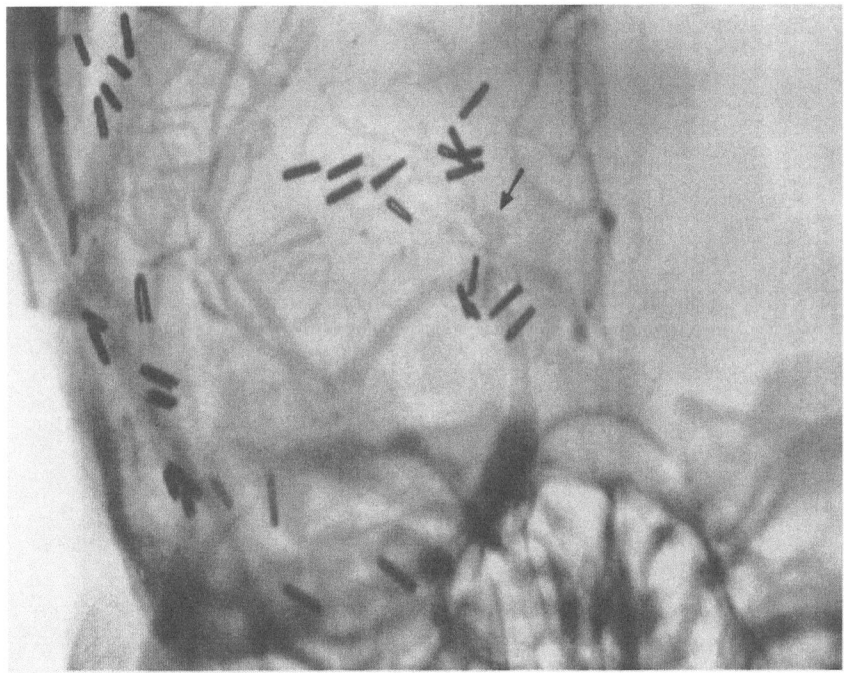

Fig. 68. See text of Fig. 67

hemiplegia persisted.  In the remaining two patients tolerating the ligation (type II), beneficial results were not observed.

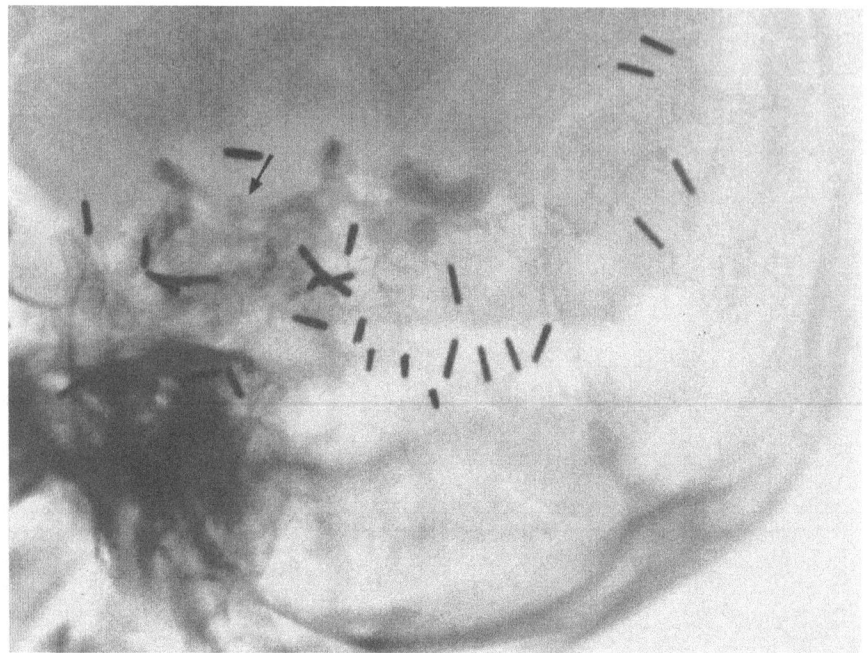

Fig. 69. Case XI. *Right vertebral angiogram* (postoperative). The lesion now is supplied by branches from
the basilar artery

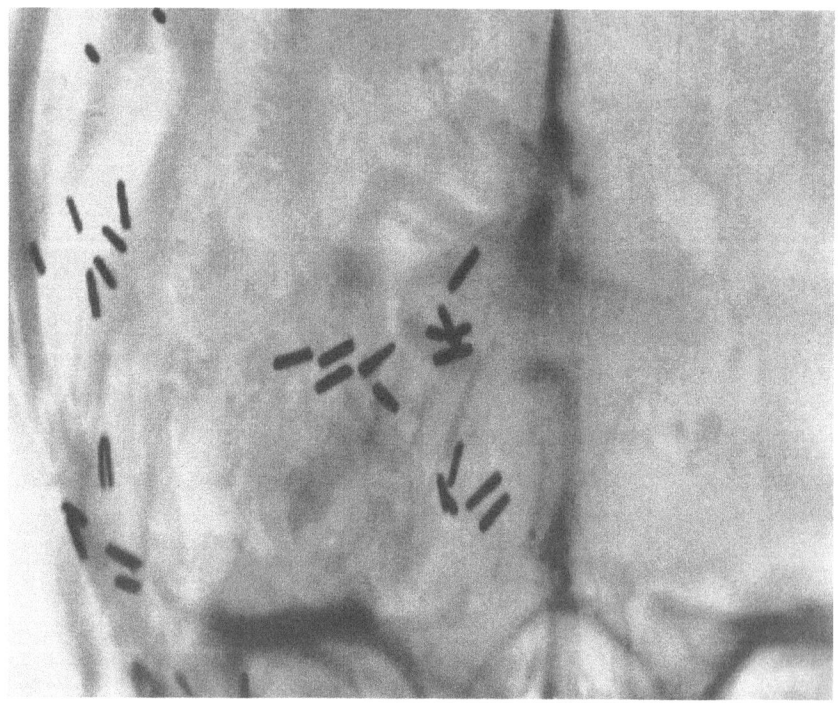

Fig. 70. See text of Fig. 69

The above procedures were given a trial at this clinic at some time in the past and
each in turn was found to be wanting. The conclusion slowly became apparent, as stated
in a previous communication[33], that *"the choice lies between removing the lesion or leaving
it alone"*.

## B. Total Excision

Total excision was performed in 80 of our 124 patients. In one patient, referred to above, ligation of the main artery proved sufficient to arrest arterial feeding, and sub-

*Table 5*

|  | 1923—1935 | 1936—1946 | 1947—1954 | Total |
|---|---|---|---|---|
| Patients . . . | 18 | 43 | 64 | 125 |
| Extirpation . . | 5 | 25 | 51 | 81 |

*Table 6*

| Causes of Operative Mortality | 1923—1955 |
|---|---|
| Uncontrollable hemorrhage . | 3 |
| Postoperative clot . . . . . | 2 |
| Cerebral edema . . . . . . | 1 |
| Meningitis . . . . . . . . | 1 |
| *Total* . . . . . . . . . | 7 |

sequent excision was not required. Table 5, divided into three chronological periods, indicates the increasing rate of surgery during the latest period.

*Mortality.* Among the 81 patients whose lesion had been excised, 7 (9 %) have expired (Table 6). During the years 1947–1950, the mortality increased somewhat when the indications for surgery were extended to larger malformations. The absence of an operative mortality since 1951 is largely attributable to the newer methods of hypotensive anesthesia now at the surgeon's disposal.

### 1. Clinical Evaluation

The 74 patients surviving the operation have been classified according to their working capacity from the information supplied from follow-up visits and yearly letters to the clinic.

*Good results.* Fifty patients (62 %) were well and able to resume work. A few had minor disabilities or an occasional epileptic attack, but, notwithstanding, were self supporting.

*Improved.* Fifteen patients (19 %) improved following surgery but retained disabilities which limited their working capacity. Although many of these disabilities were present prior to surgery, attributable for the most part to earlier intracranial hemorrhage, at times the preoperative disabilities were aggravated by surgery. One patient (A. E. J., 172/37), not listed in this category, survived surgery and enjoyed reasonably good health for several years, until the sudden onset of hemiplegia with aphasia. Angiograms demonstrated a "recurrence" of the lesion (Fig. 71, 72). Obviously the original excision had been incomplete, but since surgery had been performed before the day of the routine postoperative angiogram, both patient and surgeon were deceived in their hopes for a cure. The patient was not subjected to reoperation.

*Poor results.* In 7 patients (9 %) the results following surgery were poor, ascribable to hemiplegia in 3 and to mental derangement in 2 cases. One patient committed suicide, and one succumbed to hepatic cirrhosis. Although all of these patients were incapacitated before surgery, none derived benefit from the operative intervention.

### 2. Angiographical Evaluation

Since 1946 the clinical assessment of the operative procedure has been supplemented by postoperative angiographic studies. Where a lesion has been removed in its entirety, the prominent changes demonstrated in the angiogram following surgery include: 1, disappearance of all traces of the angiomatous vessels of the lesion; 2, return of the caliber of the outlet-inlet vessels to more normal dimension; and 3, contrast-filling of those arteries not previously visualized.

A complete postoperative angiographic study was performed in 46 of the 51 patients receiving excisional surgery during the years 1947–1955. In 40 (88 %) patients the postoperative angiogram showed complete excision of the malformation, together with the aforementioned vascular changes (Fig. 73–76). In 5 (10 %) patients the angiogram showed

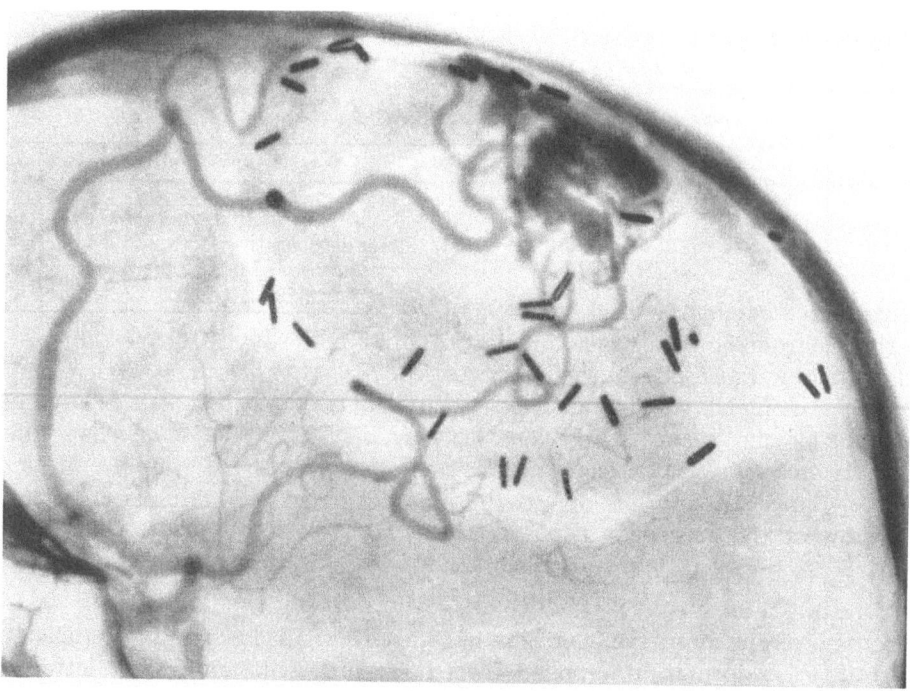

Fig. 71. E. J., 42 year old woman, 172/37. *Left internal carotid angiogram.* Lesion of the left parietal lobe, supplied by the anterior, middle and posterior cerebral arteries. Note the dilatation of the callosomarginal and posterior cerebral arteries. Drainage to the superior longitudinal sinus. An excision of the lesion, thought to be total, had been performed 14 years before

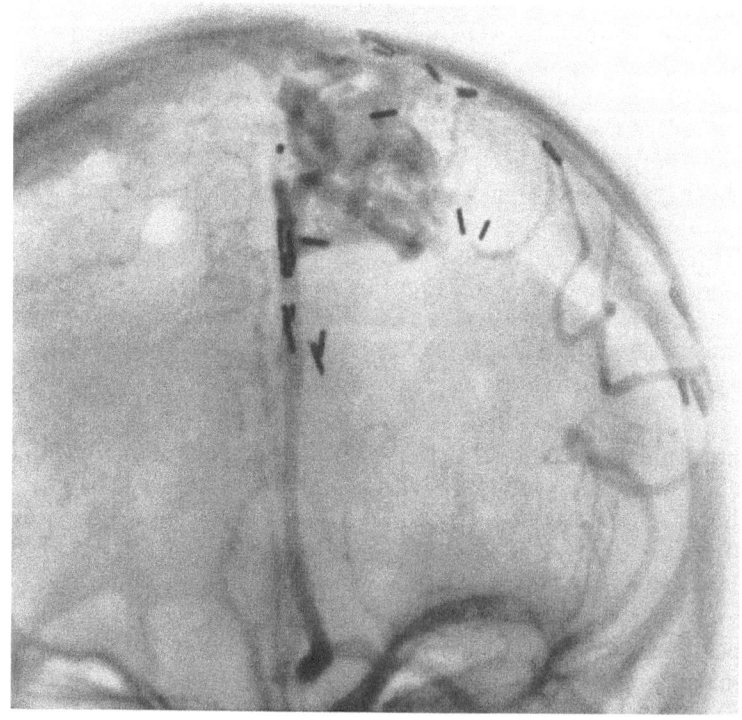

Fig. 72. See text of Fig. 71

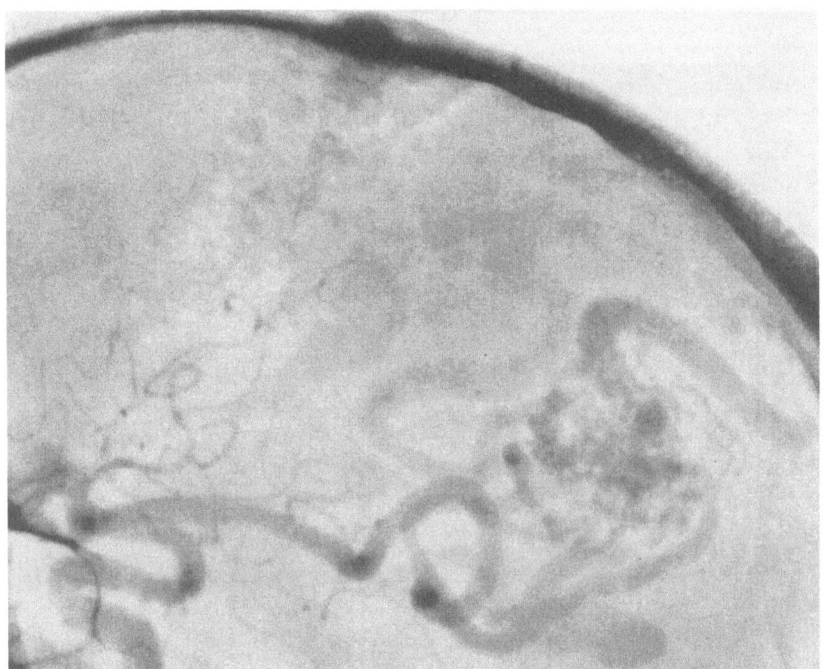

Fig. 73. A. G., 52 year old man, 420/50. Intracranial bleeding. *Left internal carotid angiogram* (preoperative). Lesion behind the sylvian fissure, supplied by two branches of the middle cerebral artery. Note the poor visualization of the vessels which do not supply the malformation. Drainage to the sagittal sinus

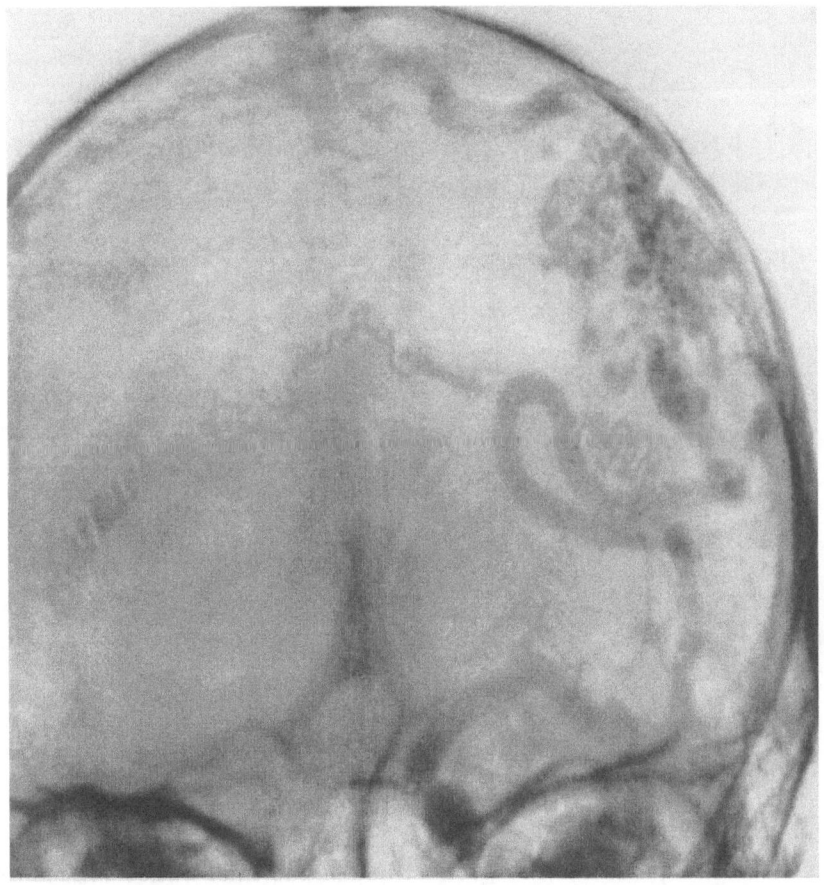

Fig. 74. See text of Fig. 73

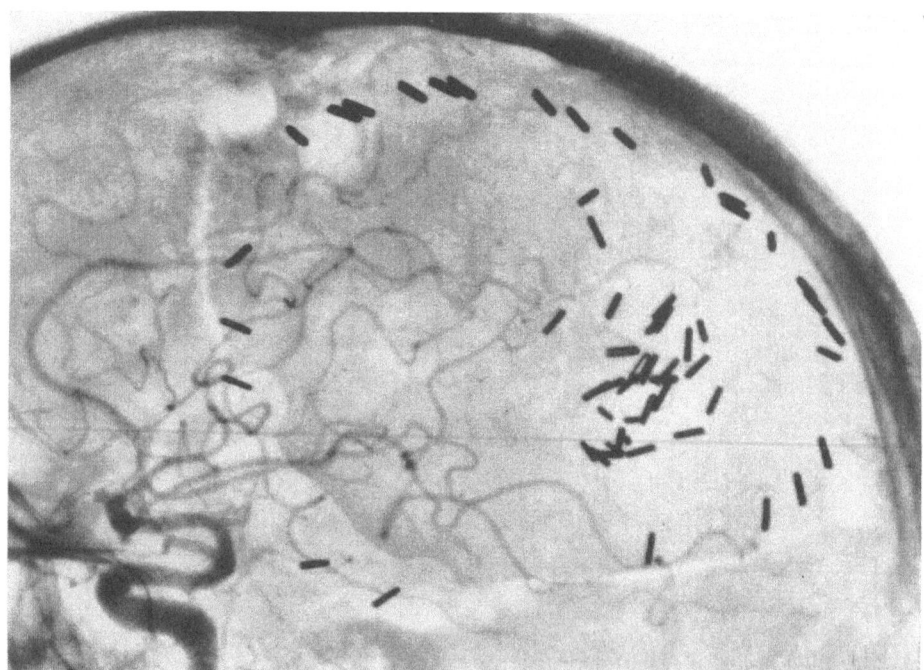

Fig. 75. A. G., 52 year old man, 420/50. *Left internal carotid angiogram* (postoperative). Malformation extirpated. Normal caliber of the cerebral vessels. The patient is now fully employed

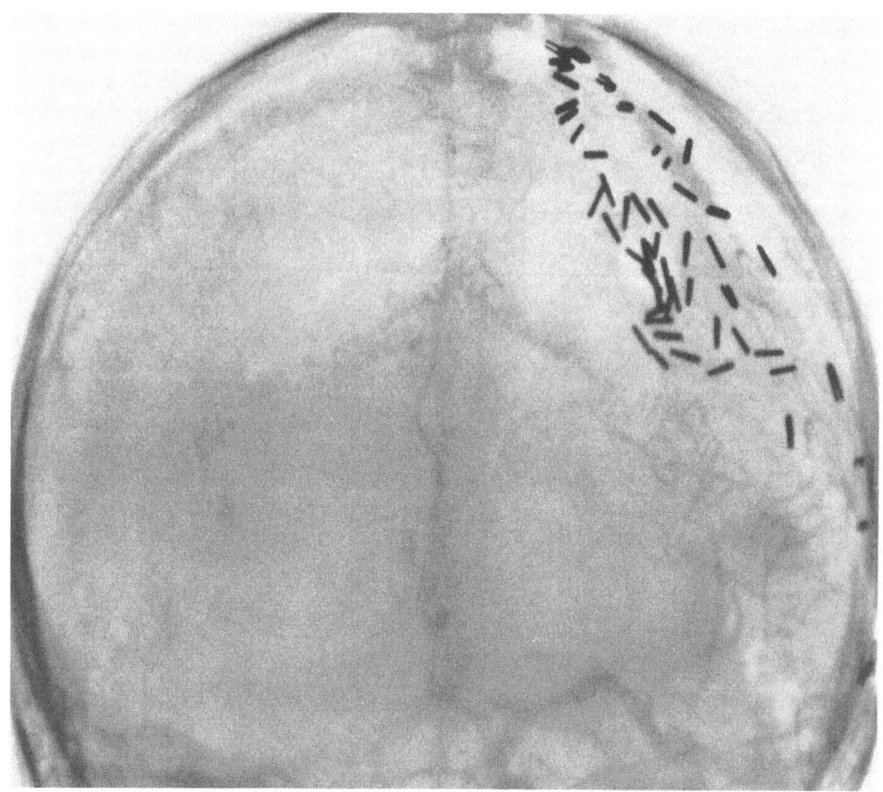

Fig. 76. See text of Fig. 75

the excision to be not quite complete, for, although the inlet-outlet vessels returned to normal size and arteries could be demonstrated which had not previously been visualized in the preoperative angiogram, nevertheless vestiges of the angioma could be discerned, often only after close scrutiny of the X-ray film (Fig. 77–80).

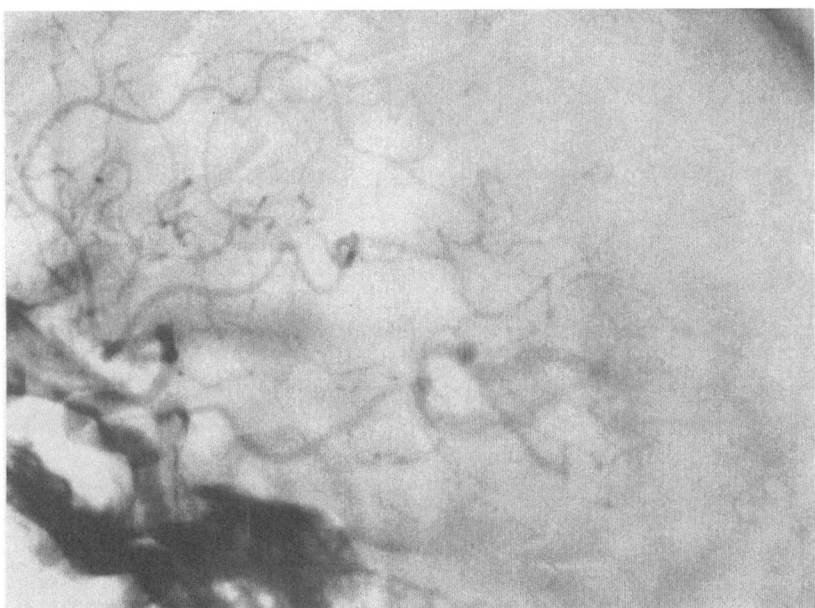

Fig. 77. H. M. R., 34 year old woman, 230/54. Intracranial bleeding 1 month before admission. *Right interna carotid angiogram* (preoperative). Lesion of the occipito-temporal region, supplied by the posterior cerebra artery. Drainage into the sinus rectus. Note the displacement of the pericallosal artery to the left

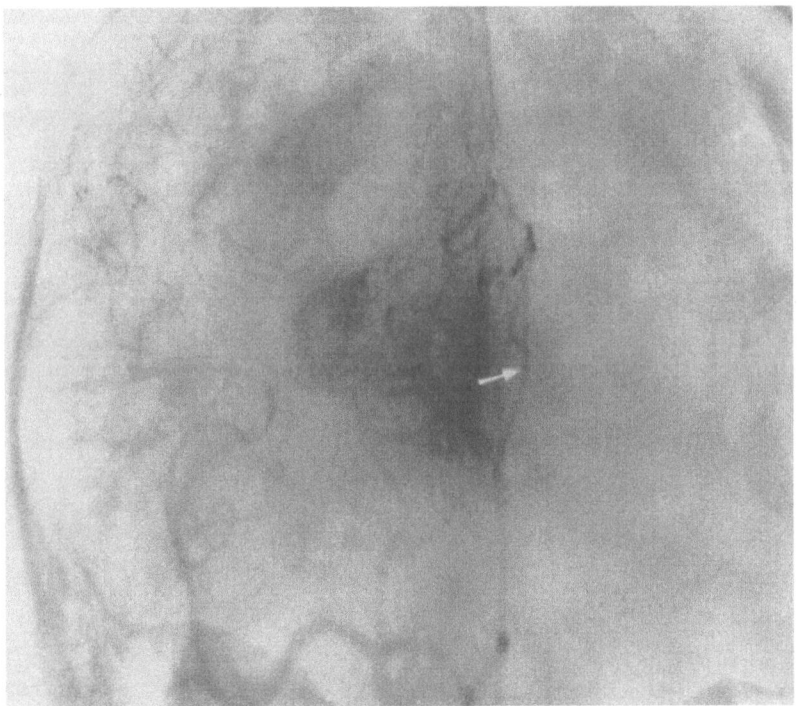

Fig. 78. See text of Fig. 77

The effect of small amounts of residual angiomatous tissue on the ultimate prognosis of the lesion is at present difficult to predict. We have not thus far observed either a

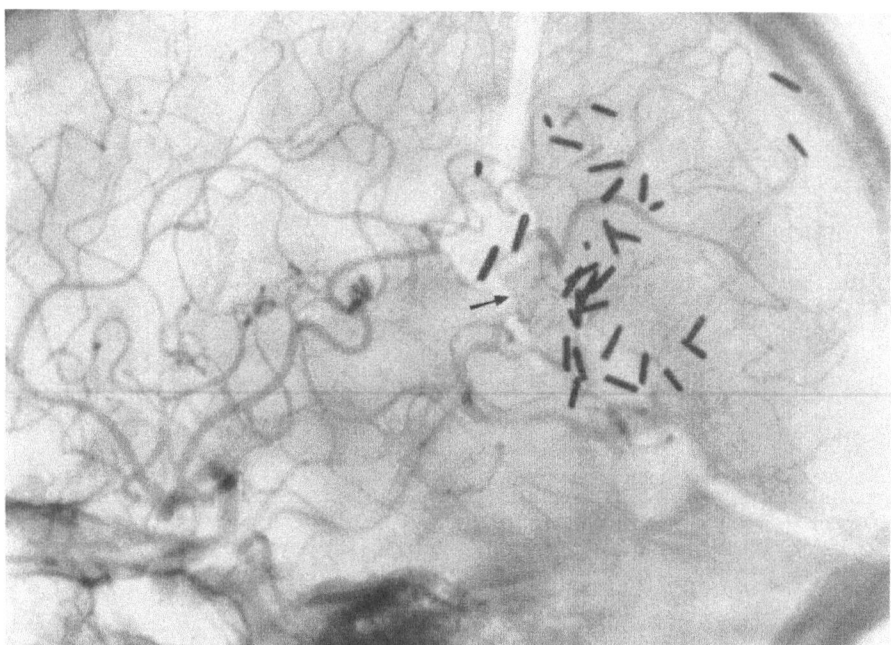

Fig. 79. H. M. R., 34 year old woman, 230/54. *Right internal carotid angiogram* (postoperative). The bulk of the lesion has been removed, but a small residuum remains. Visual field defects and hemiplegia, present since the initial vascular insult, abated considerably within 3 months after surgery

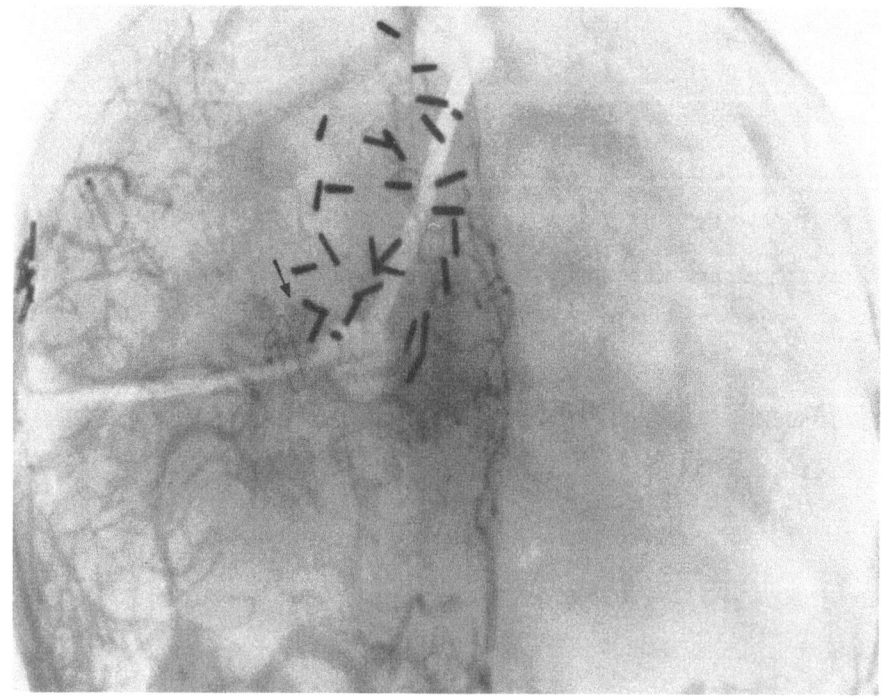

Fig. 80. See text of Fig. 79

recurrence of bleeding or subsequent clinical deterioration in any of these 5 patients. Indeed, the clinical result was often as gratifying as if the aneurysm had been excised

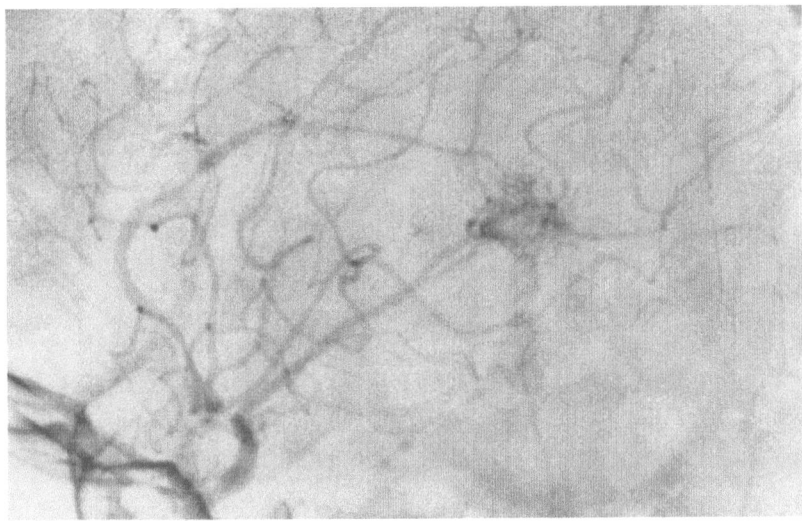

Fig. 81. A. M. B., 11 year old girl, 416/54. *Left common carotid angiography* (preoperative). A walnut-size lesion in the left parietal hemisphere, located at the level of the splenium of the corpus callosum. The pericallosal artery supplies the lesion. Drainage to the great vein of Galen

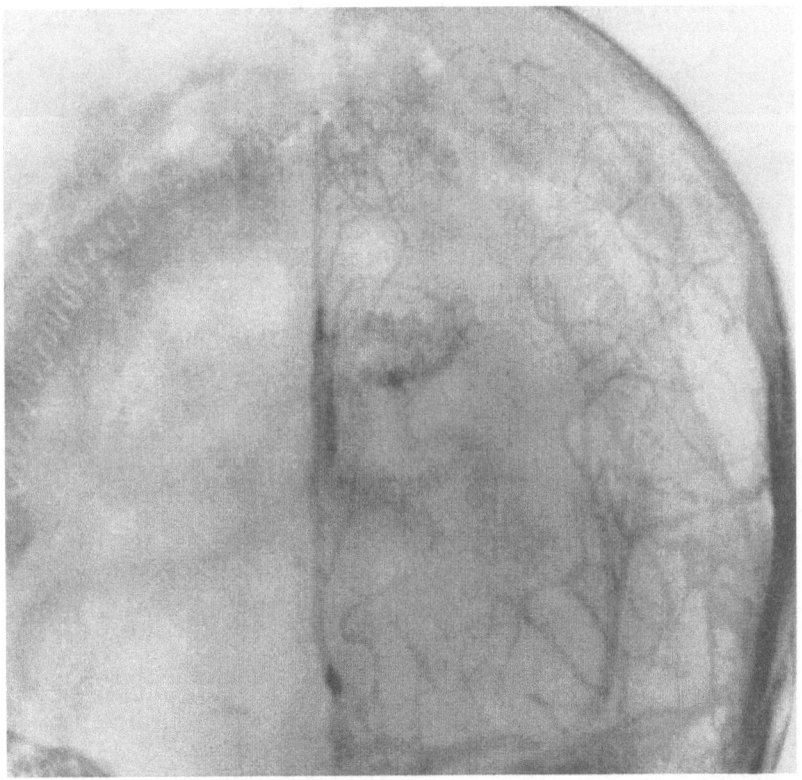

Fig. 82. See text of Fig. 81

in its entirety. This observation has fortified our conviction that these patients may be provisionally characterised as deriving beneficial results from surgery, but it must be clearly understood that this opinion may have to be revised at a later date.

In one case (2 %) the operative result was unquestionably unsatisfactory (Fig. 81–84). The patient (A. M. B., 416/54), a 10 year old girl, whose first admission was preceded by

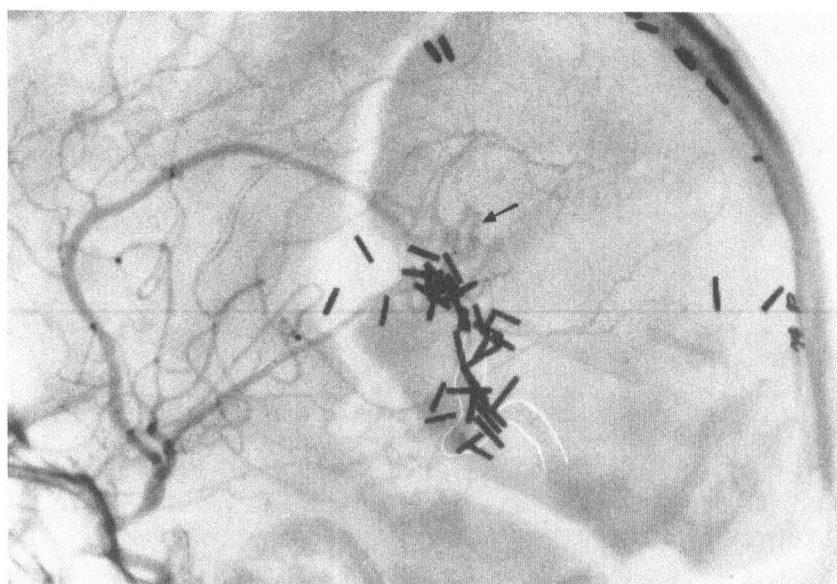

Fig. 83. A. M. B., 11 year old girl, 416/54. *Left common carotid angiography* (postoperative). The middle portion of the arteriovenous aneurysm can still be visualized. See text

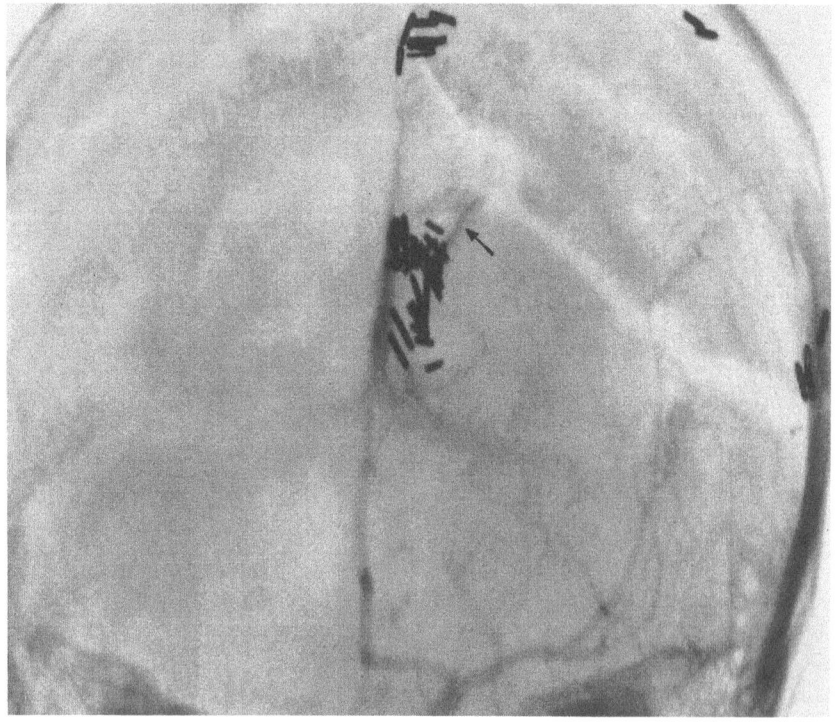

Fig. 84. See text of Fig. 83

intracranial hemorrhage, experienced another vascular insult 5 months after an incomplete surgical extirpation. Reoperation was performed as an emergency measure, but, obscured by the hemorrhagic changes, the residual angiomatous tissue escaped detection. Further operative intervention was planned for the future, but strained relations with the girl's parents (the patient was the only child of a physician) denied us this opportunity. The remaining tissue has since been removed by Dr. G. NORLÉN.

## 3. Prognosis with Surgery

Several considerations must be reviewed before a value for the unfavorable results can be assigned. *1. Incomplete extirpation* (12%). We assume that the value derived from the angiographic series also obtains for the patients who underwent surgery prior to 1946. Should all of the extirpations classified as incomplete (or almost complete) be considered unsatisfactory? Admittedly, the evidence adduced earlier indicates that smaller lesions are more disposed to bleeding than the larger malformations, but the fact remains that only 1 (2%) of the patients in this category has had an unfavorable clinical outcome subsequent to the procedure. *2. Poor clinical results* (9%). Not every patient evaluated as a poor clinical result after complete excision can attribute this outcome to surgery. Half of these patients had irreversible disabilities resulting from hemorrhagic destruction of brain tissue. The remainder (4%) were somewhat further incapacitated by surgery, but, more important, the ever-present threat of further hemorrhage had been removed. *3. Surgical mortality* (9%). This value must be accepted without qualification for the series as a whole. There has not been, as previously noted, a single operative death from 1951–1956.

The *maximum total unfavorable results* of extirpation for the entire series (1923–1955) is *30%*, the sum of the unfavorable clinical results (9%), the incomplete excisions (12%) and the operative mortality (9%). The *minimum unfavorable results* is *15%*, the sum of the number who were clinically harmed by surgery (4%), the operative mortality (9%), and the number with incomplete excision who later experienced untoward clinical sequelae (2%).

The practical surgeon is more concerned with statistics which reflect the prevailing situation rather than those which pertain to the decades antedating hypotension anesthesia and the modern supplemental techniques.

As matters stand today, if a patient with a lesion which meets the standards of operability undergoes extirpative surgery with hypotension anesthesia, the probabilities for surgical mortality should be negligible. About *85%* of these patients will obtain clinical improvement, complete excision of the angiomatous tissue and freedom from future hemorrhage. Few of the remaining *15%* will experience detrimental results. 1. The lesion may be totally removed, but the preexisting symptoms will be somewhat aggravated. This is a small premium to pay for insurance against further vascular insult. 2. A small residual of angiomatous tissue may remain, but the symptoms will be improved. 3. Residual tissue may persist with aggravation of the preoperative symptoms. This last group may require further surgical intervention at a later date.

## C. Palliative Treatment

The end results in the remaining 44 cases can best be appreciated by treating them as a single group, since in none was a definitive procedure performed.

*Mortality.* Of the 14 patients who have died, 2 succumbed to palliative surgery. One patient, the first in this series (1923), died following a first-stage cerebellar exploration. The second died from hemorrhage during an attempt to ligate surface vessels. Twelve other patients in the palliative group have died, presumably from intracranial hemorrhage. In some instances, the causes of death were established by post-mortem examination, while at other times, the continuity of complaints up to the time of expiration leaves little doubt as to the cause of death.

*Survival.* Of the 30 patients who survived, 16 are alive and well, except for such minor inconveniences as an occasional epileptic attack. Nine patients are partially invalided (epilepsy 4, sequelae of intracerebral hemorrhage 5) while in 7 patients the incapacitation is complete (hemiplegia 5, blindness 1, mental derangement 1, cerebellar disturbances 1).

*Posterior fossa lesions.* The early expectations of this clinic with regard to malformations of the posterior fossa have not been realised. To date only one case has been encountered where the anatomical situation of the lesion appeared to permit complete extirpation. More often the extension of the lesion to the region of the incisura, the contralateral side or the ventricle presented difficulties which were surgically insurmountable.

*Table 7*

| | Palliative Treatment |
|---|---|
| Exploration . . . . . | 10 |
| Decompression . . . . | 2 |
| Afferent, surface, or limited ligations . . | 6 |
| No surgery . . . . . | 18 |
| Carotid artery ligation | 6 |
| *Total* . . . . . . . | 44 |

Early in this series a partially calcified lesion (Fig. 55) was "accidentally" encountered in the left cerebellar lobe of a 35 year old male (J. R. V. W., 1925/30). The lesion, together with a large part of the cerebellar lobe, was removed during an 8 hour procedure. Following surgery, the patient exhibited marked cerebellar derangement for several years, but thereafter the symptoms subsided and at the end of 25 years he is alive and fully-employed. Admittedly, angiographic follow-up was not performed at this early date, but the radical extent of the excision and the subsequent clinical course of the patient provisionally suggest that the excision had been complete.

The remaining 8 patients received what must be adjudged palliative treatment. In two cases, where the lesion was situated in the cerebello-pontine angle, either tractotomy or Fifth Nerve section was performed for intractable pain. Two lesions received no surgical treatment because of the prohibitive appearance of the angiograms. Three patients were explored, and after extensive mobilization, involving in one case a concomitant occipital flap, the futility of attempting resection became apparent. In the ninth patient an extirpation of the angiomatous tissue was attempted, but at the end of the procedure it could be seen that the excision had been incomplete.

### Case XII

Man, age 22. Arteriovenous aneurysm of the posterior fossa. Incomplete excision. Partial working capacity.

*Admission* (Oct. 30, 1949). H. E. E., 836/49, a 22 year old male whose illness began abruptly 6 months earlier with dizziness, vomiting, diarrhea, occipital headache, ataxia in the right extremities and diplopia. *Examination.* A right-sided facial paresis, nystagmus on right gaze, dysmetria and ataxia of the right extremities, unstable gait and weakness of the left abdominal reflexes. Skull films and angiography of both the right external and right internal carotid arteries were negative. *Right vertebral angiography.* In the posterior fossa an arteriovenous aneurysm, situated behind the foramen magnum, could be seen to extend to the midline. Dilatation of the vertebral artery was present. The aneurysm was drained by veins coursing behind the third ventricle (Fig. 85, 86). *Operation* (Nov. 3, 1949). Bilateral suboccipital craniectomy. The aneurysm was not apparent at first inspection, but after the tonsil had been raised, a dilated posterior inferior cerebellar artery could be seen. The vessel was clipped and divided. Thereafter, the dissection proceded from the vermis to the upper lateral half of the right cerebellar lobe and from the surface to the ventral aspect of the cerebellar lobe. At the end of the procedure an arterialized vein was observed to course on the left lobe, suggesting that angiomatous tissue had been left behind, probably involving the vessels which had been observed to pass behind the ventricle. Further manipulation was thought to be unwise. The excised specimen weighed 38 grams. *Postoperative course.* Convalescence was complicated by meningitis, which slowly resolved under chemotherapeutic treatment. *Postoperative angiography* of the right vertebral artery demonstrated some diminution of the bulk of the lesion, but a significantly large residuum could still be seen (Fig. 87, 88). *Follow-up.* The dysmetria and ataxia diminished following surgery, but the patient, a music director, was at last reports still troubled with clumbsiness, especially when writing or playing the piano.

**Comment.** The patient was a 22 year old youth who enjoyed the best of health before his illness. Whether or not a more extensive excision would have removed the lesion *in toto*, is a matter for speculation, but certainly there is little doubt that at best the patient would have been left a bed-ridden invalid. Under these circumstances it was felt that further intervention was not in his immediate interest.

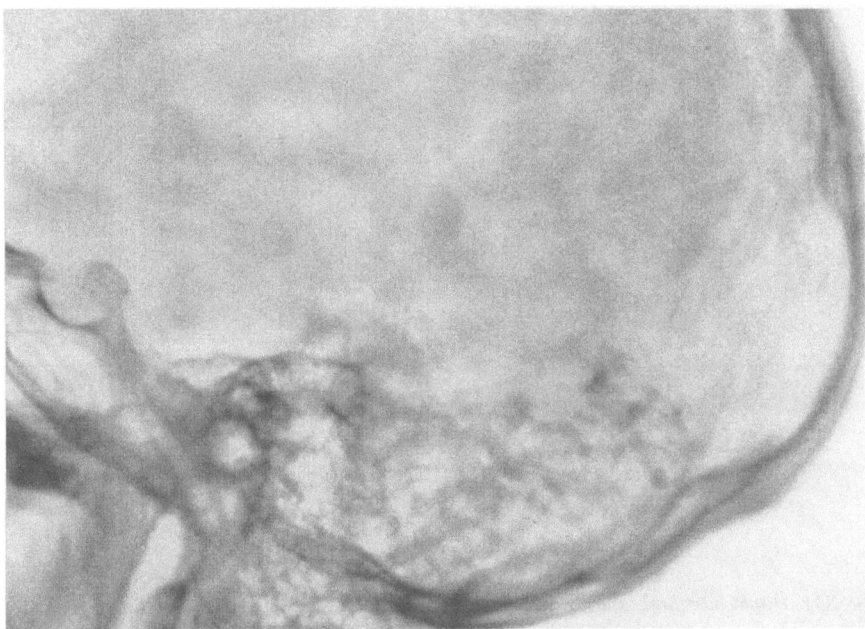

Fig. 85. Case XII. *Right vertebral angiogram* (preoperative). A posterior fossa lesion situated behind the foramen magnum. The vertebral artery is enlarged and contributes to the malformation. The contrast chiefly passes to the aneurysm while the basilar artery receives only a fragment of the supply. Drainage to veins coursing behind the third ventricle. The right internal carotid arteriogram showed no abnormality

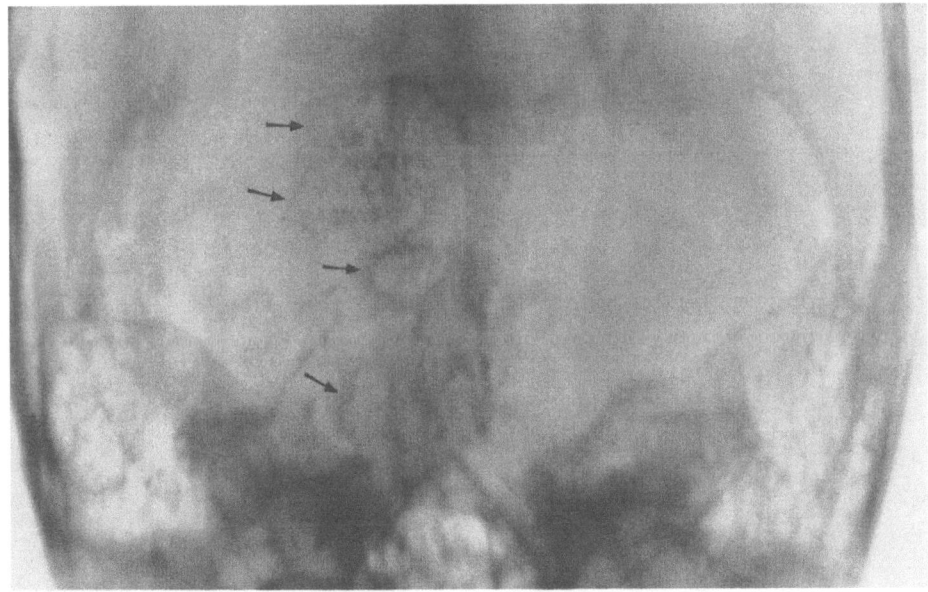

Fig. 86. See text of Fig. 85

Notwithstanding our own dismal experiences with the posterior fossa lesion, there is no justification for abiding pessimism, and the authors strongly repudiate the suggestion that this lesion must be written off the surgical books. All of the above-mentioned procedures were performed before the advent of hypotensive anesthesia. Furthermore, since the number of cerebellar malformations in this series is very small, it cannot be excluded

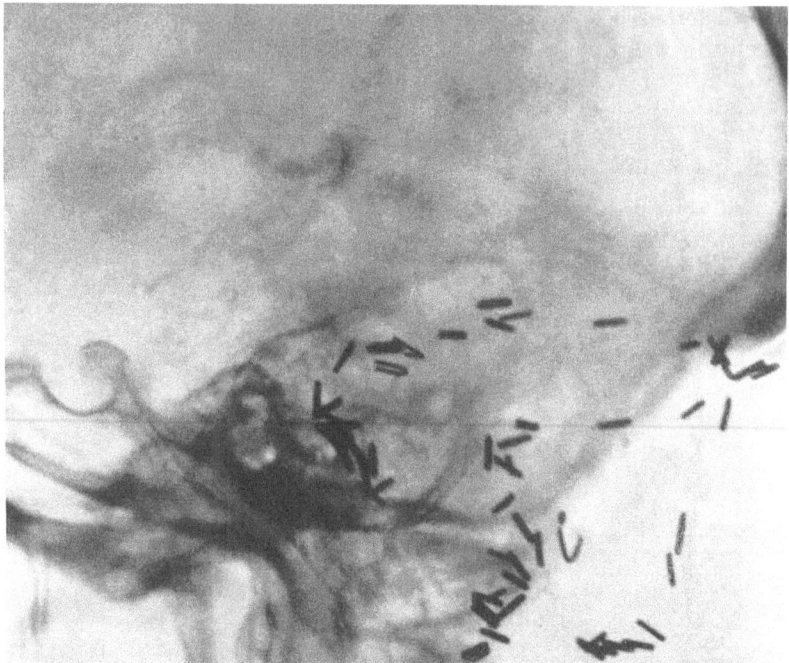

Fig. 87. Case XII. *Right vertebral angiogram* (postoperative). A large part of the aneurysm has been left
behind

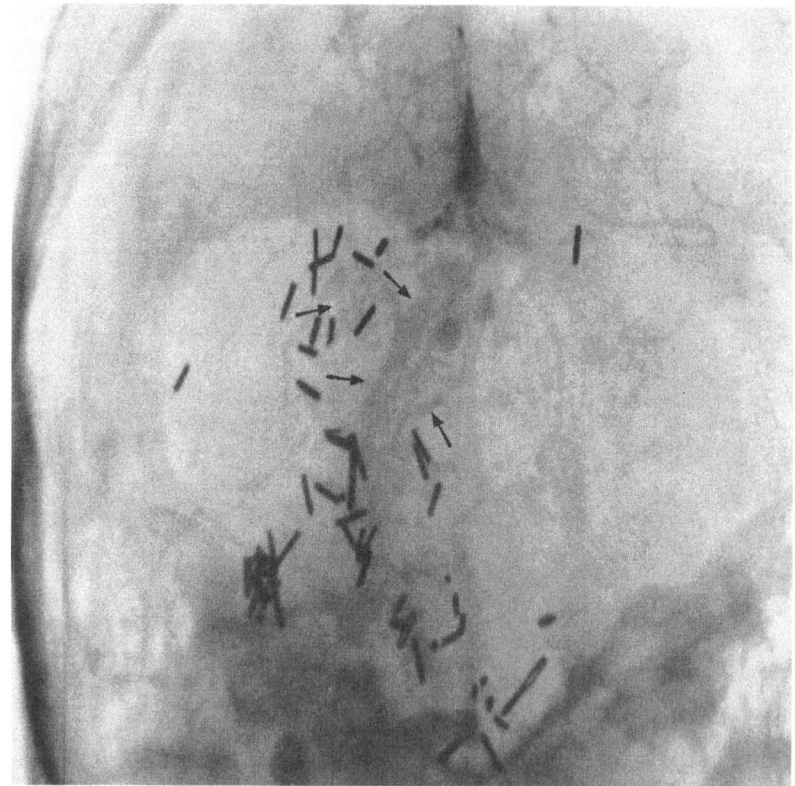

Fig. 88. See text of Fig. 87

that the next lesion to appear will, from its size and situation, allow a complete excision
without producing extensive derangement of cerebellar or brain stem function. Lastly,

hypothermia offers considerable promise in the surgery of this lesion and may well provide the answer to our present perplexities.

*Non-surgical patients.* Although it is our belief that the prognosis for patients who have not undergone palliative surgical procedures will in no appreciable way differ from the prognosis of patients receiving palliative surgery, it may be well to consider the untreated cases as a separate entity, lest the objection be raised that the results in the palliative group as a whole were excessively poor because many of the lesions had been

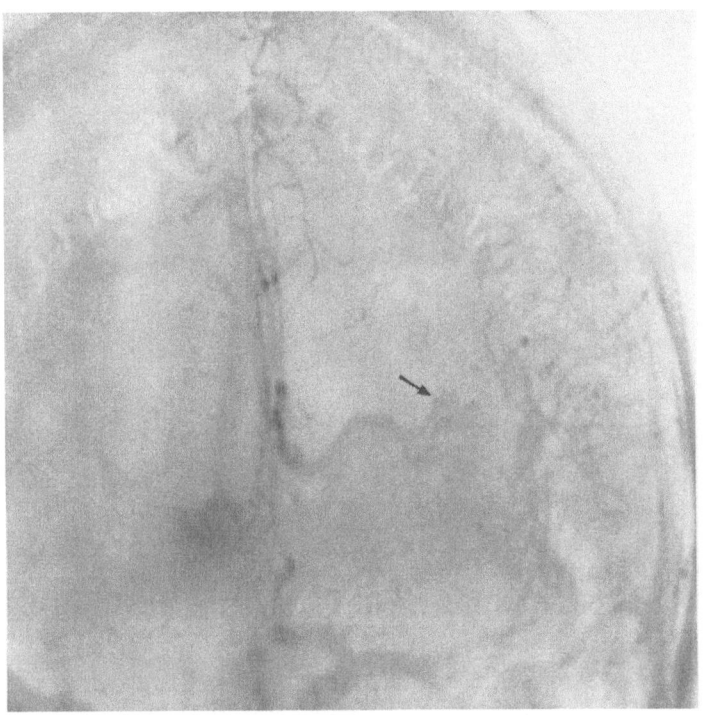

Fig. 89. K. V. A., 34 year old man, 977/51. Intracranial hemorrhage 2 years earlier. *Left internal carotid angiogram.* A small lesion of the frontal lobe, supplied by a vessel from the sylvian group. Drainage medially and inferiorly to the vena magna and sinus rectus

ineffectually tampered with by the surgeon. Dr. O. Höök has called our attention to this criticism.

Of the 18 patients who were "left alone", 6 (33%) have died within 7 years following discharge from the clinic. The follow-up reports prior to expiration permit the presumption of death from hemorrhage. Among the 12 patients who survive today, 4 (22%) are without complaints. Despite the fact that several of these patients have, without the approval of the clinic, returned to their previous employment, we hesitate to use the term "fully employable", for no patient with this lesion can safely undertake strenuous work. Four (22%) of the surviving patients in this category are partially invalided, and 4 (22%) are completely incapacitated.

To present the results of conservative treatment in their best light, the outcome of 55% of the patients leaving the hospital without surgery will be *unsatisfactory* within a period of 7 years. As the interval of follow-up increases, the results become proportiately poorer. These statistics, of course, are based upon the inoperable lesions. It is our belief, however, that the results of conservative treatment of the operable lesions would not be appreciably different.

*A comparison of the results of radical and conservative treatment presented in this study provides more than ample justification for recommending surgery for a lesion which meets the criterion of operability.*

# D. Operability

**Indications.** The excision of extracranial lesions presents no great risk to the patient. When the lesion is situated in the region of skeletal musculature, it is prudent to warn

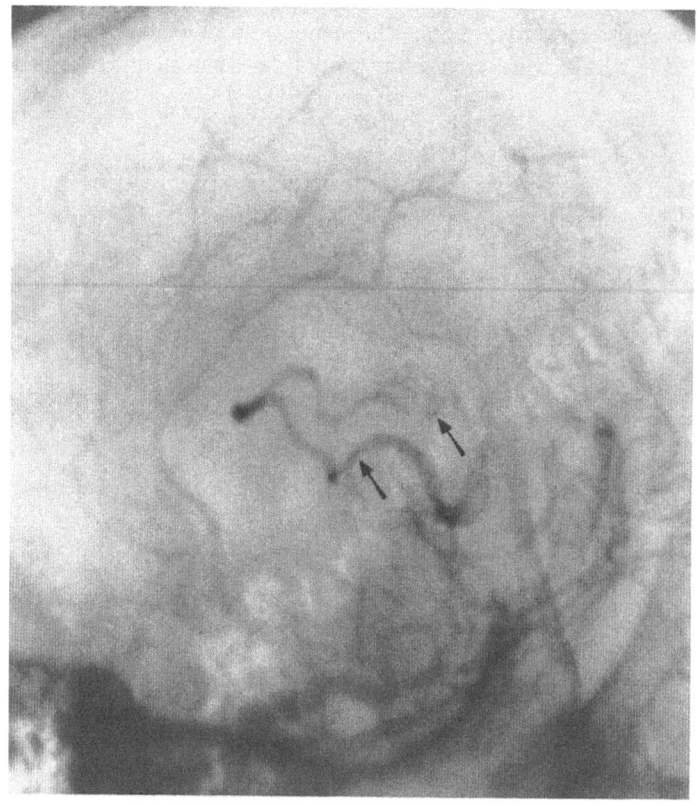

Fig. 90. An oblique view of the preceding angiogram

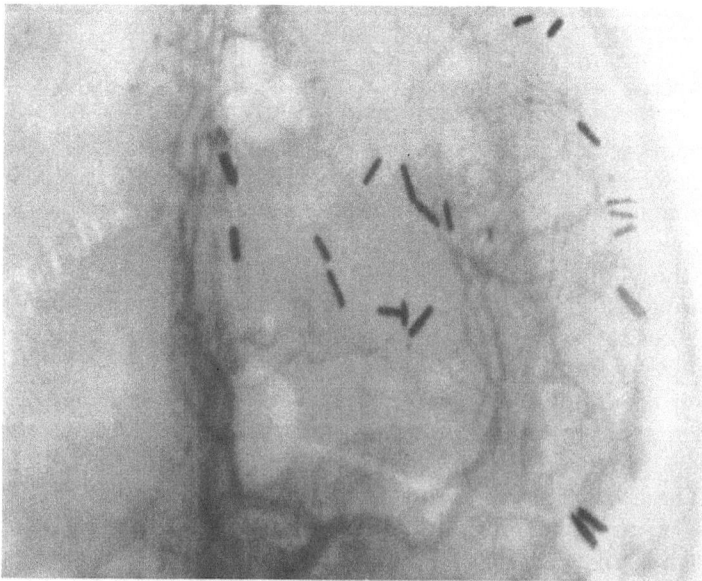

Fig. 91. K. V. A., 34 year old man, 977/51. *Left internal carotid angiogram* (postoperative). The lesion is no longer present. Patient's clinical course has improved somewhat, but many of the symptoms caused by the previous hemorrhage remain

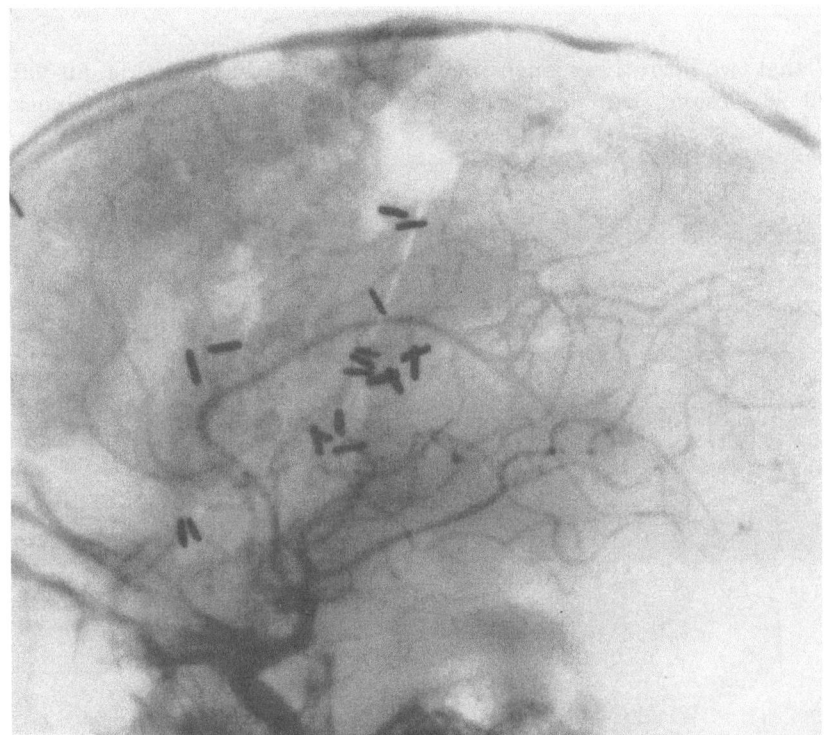

Fig. 92. See text of Fig. 91

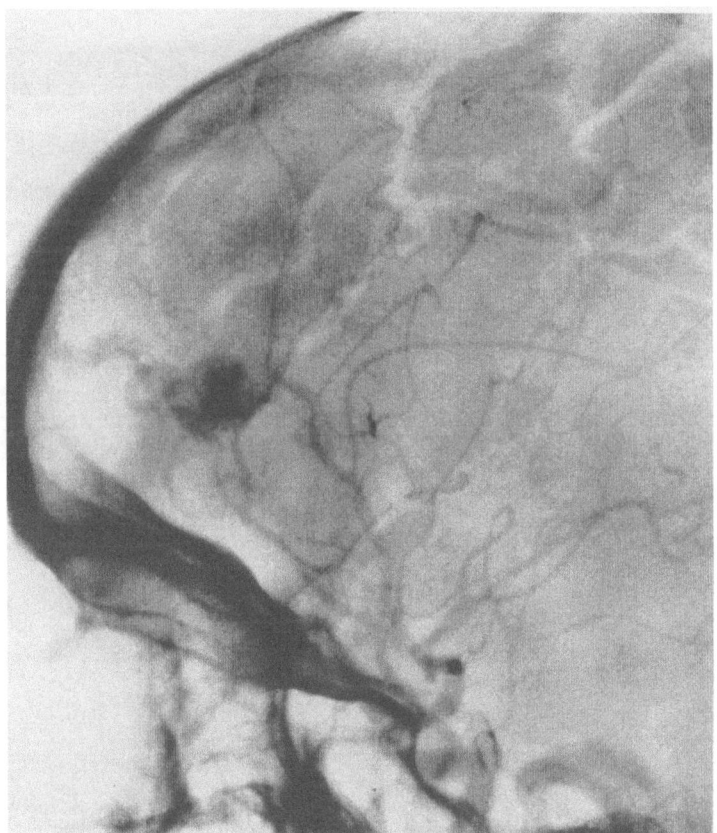

Fig. 93. A. K., 21 year old man, 650/53. Intracranial bleeding 3 months earlier. *Left common carotid angiography* (preoperative). A frontal pole lesion, fed by the frontopolar branch of the left anterior cerebral artery, which courses in the interhemispheral fissure. Drainage to the longitudinal sinus

the patient that two operative séances may be required. Frontal (Fig. 89–92) and occipital (Fig. 93–99) lesions can be removed with relative safety, since block dissection

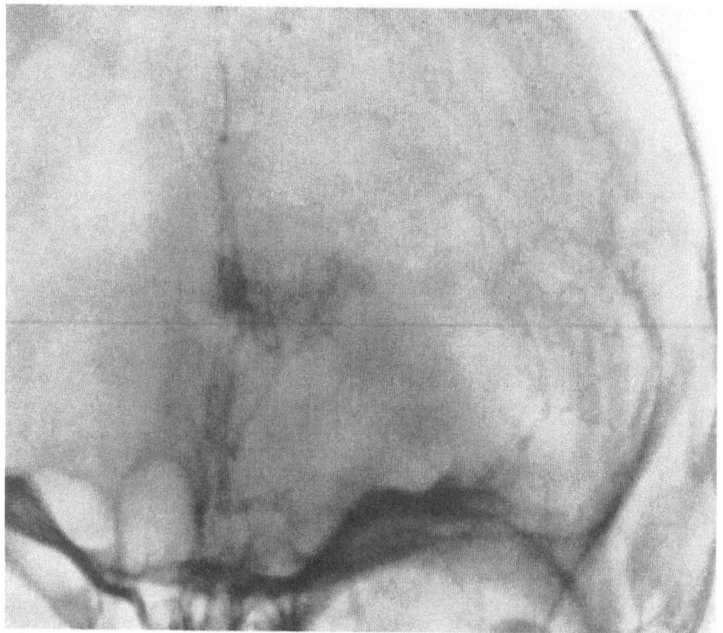

Fig. 94. See text of Fig. 93

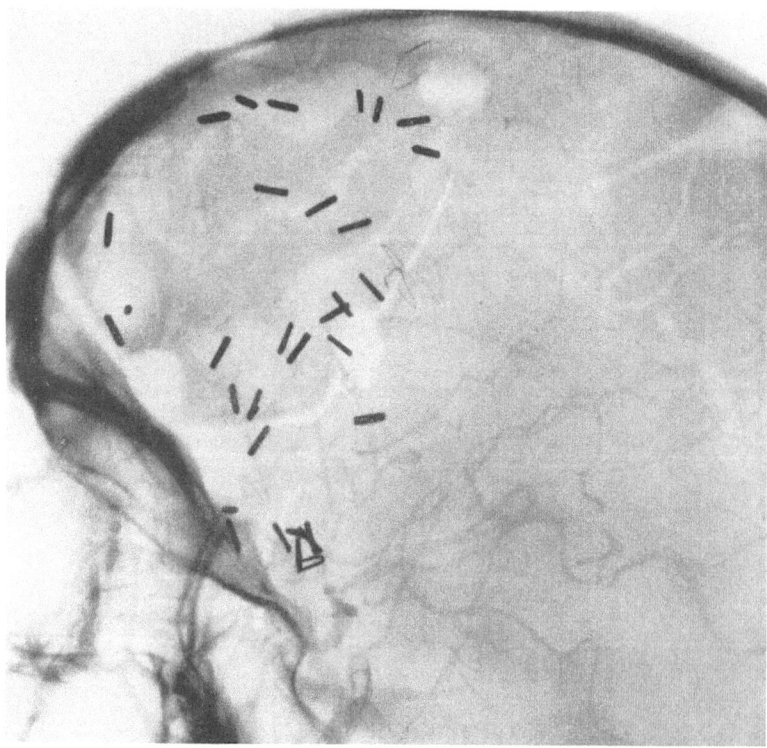

Fig. 95. A. K., 21 year old man, 650/53. *Left common carotid angiography* (postoperative). The malformation has been removed. The patient is fully employable

constitutes little danger for permanent disability. Intracranial hemorrhage should not, *per se,* be a deterrent for prompt surgery when the condition of the patient might other-

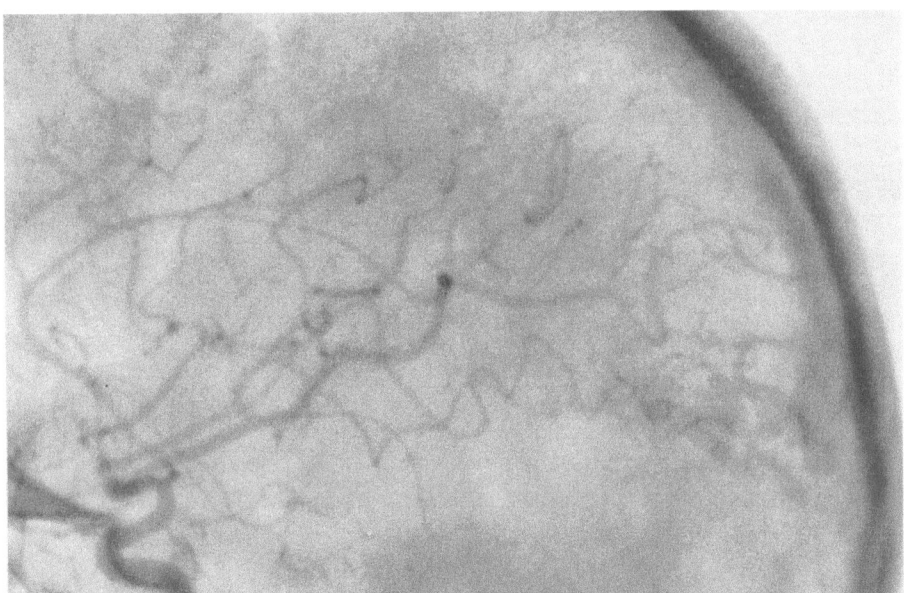

Fig. 96. E. I. S., 20 year old woman, 852/52. Intracranial bleeding 1 month earlier. *Right common carotid angiogram* (preoperative). A right occipital lesion, supplied by the enlarged posterior temporal artery. Drainage to the superior longitudinal sinus. Note that the posterior cerebral artery does not visualize

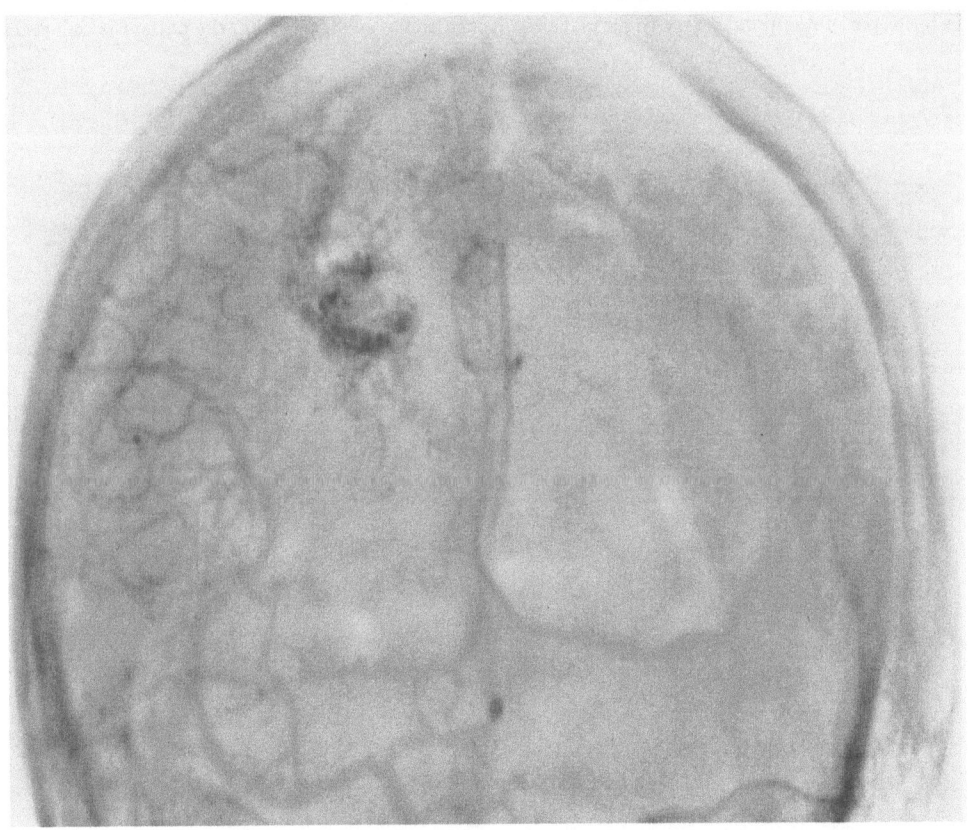

Fig. 97. See text of Fig. 96

wise be hazarded by delay, for the removal of the clot will often expose a partially mobilized lesion.

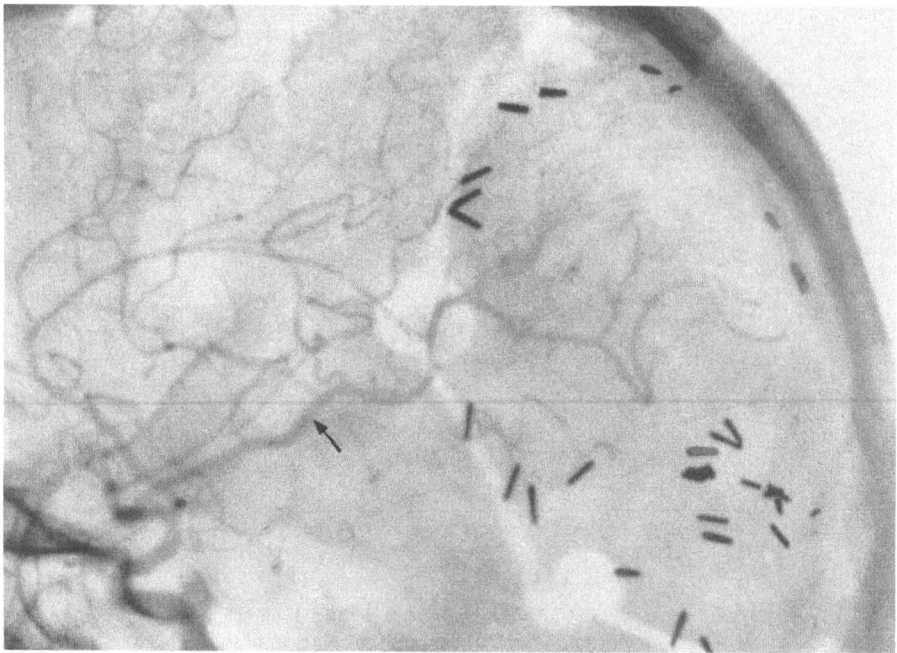

Fig. 98. E. T. S., 20 year old woman, 852/52. *Right common carotid angiogram* (postoperative). The lesion has been removed. Better contrast filling of the pericallosal artery can be seen. Note the return of the caliber of the posterior parietotemporal artery to normal size. Apart from a small scotoma, the patient is well and fully employed

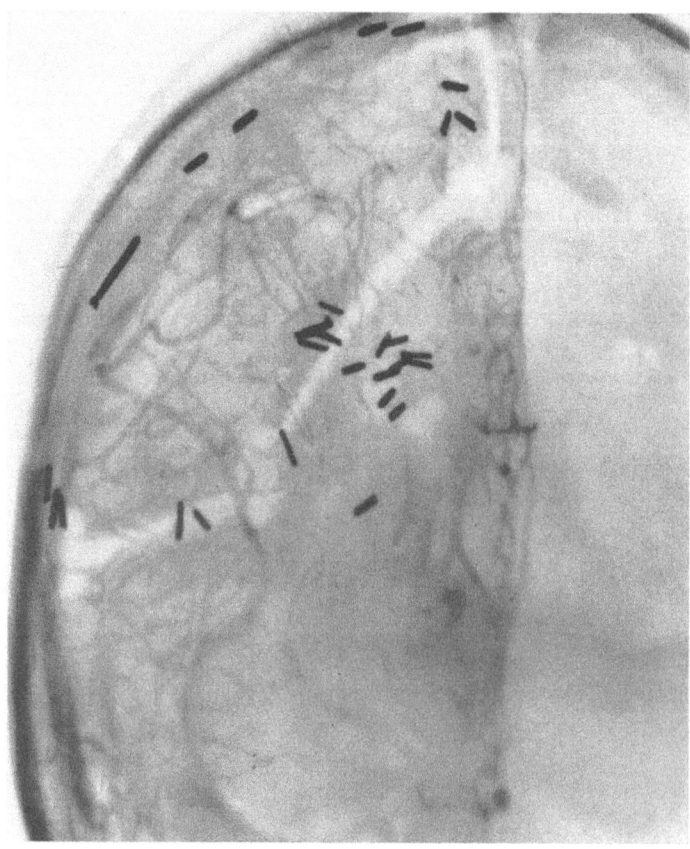

Fig. 99. See text of Fig. 98

Lesions of the motor cortex offer the greatest difficulties and invoke the need for sound surgical judgement. Small malformations can at times be cautiously attacked

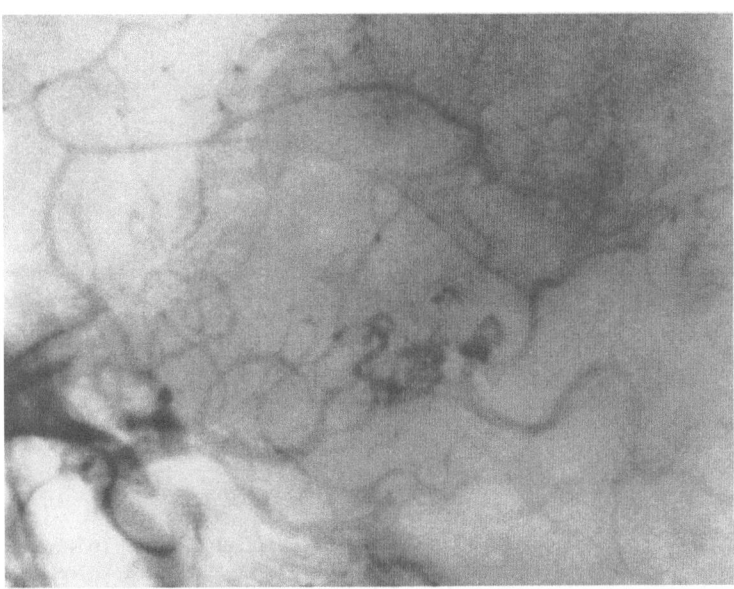

Fig. 100. S. A. K., 34 year old woman, 249/53. Intracranial bleeding 1 month earlier. *Right internal carotid angiography* (preoperative). A dilated branch of the middle cerebral artery supplies a bean-sized lesion. Drainage to the left internal cerebral vein. The aneurysm is of unusually small size

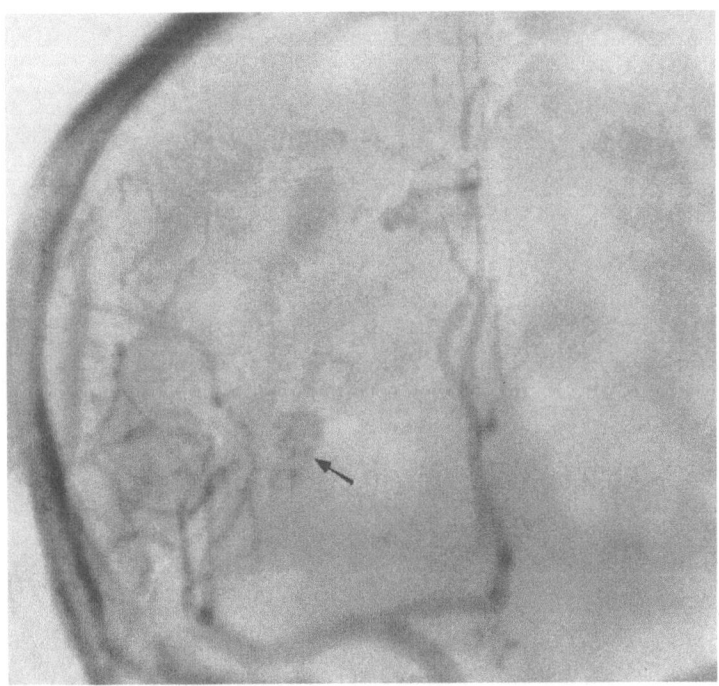

Fig. 101. See text of Fig. 100

(Fig. 100–103), and, in the region of the Sylvian fissure, lesions can often be mobilized by opening the fissure to visualize the feeding vessels (Fig. 104, 105). In considering the extirpation of the lesions situated in precarious areas (Fig. 106–109), one must remember that half of the patients have had a history of intracranial bleeding prior to

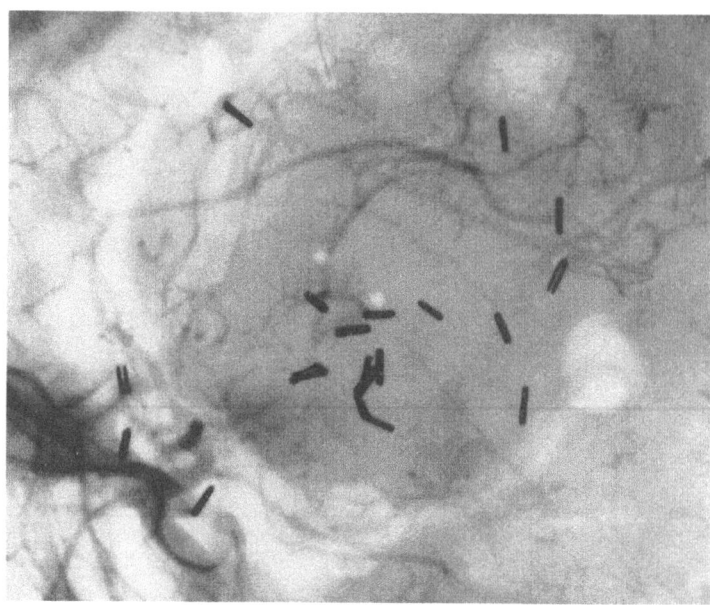

Fig. 102. S. A. K., 34 year old woman, 249/53. *Right internal carotid angiography* (postoperative). The arterio-venous aneurysm is not visualized and in its place can be seen a large area of relatively poor vascular filling. This often occurs when the postoperative angiogram is performed soon after surgery. Better visualization of the area can be obtained at a later date. Apart from the hemiparesis from the previous hemorrhage, the patient is well

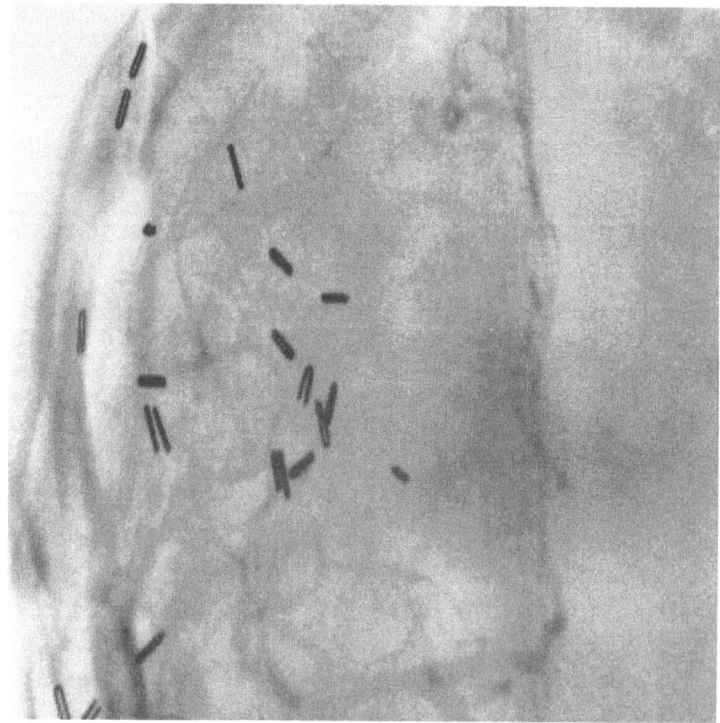

Fig. 103. See text of Fig. 102

admission and those who leave the hospital without surgery can await little more than the dismal prospects of incapacitation, mental derangement and finally death from hemorrhage. On the other hand, the patient not undergoing surgery may survive for

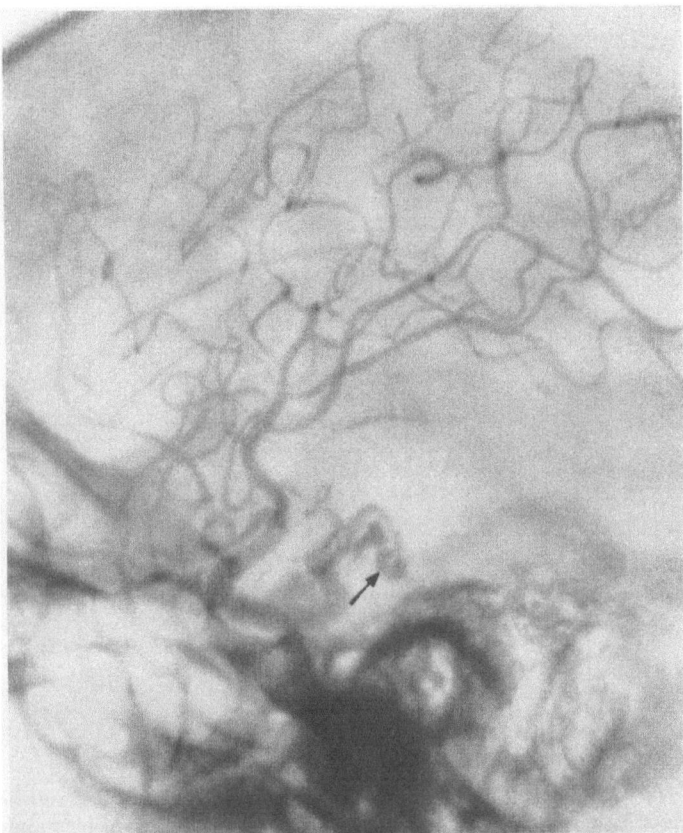

Fig. 104. H. M. S., 34 year old woman, 236/53. Intracranial hemorrhage 1 month earlier. *Left internal carotid angiography* (preoperative). The lesion is located in the anterior basal part of the left temporal lobe at the level of the posterior wall of the sella. The blood supply is probably derived from branches of the middle cerebral artery near its division in the sylvian fissure

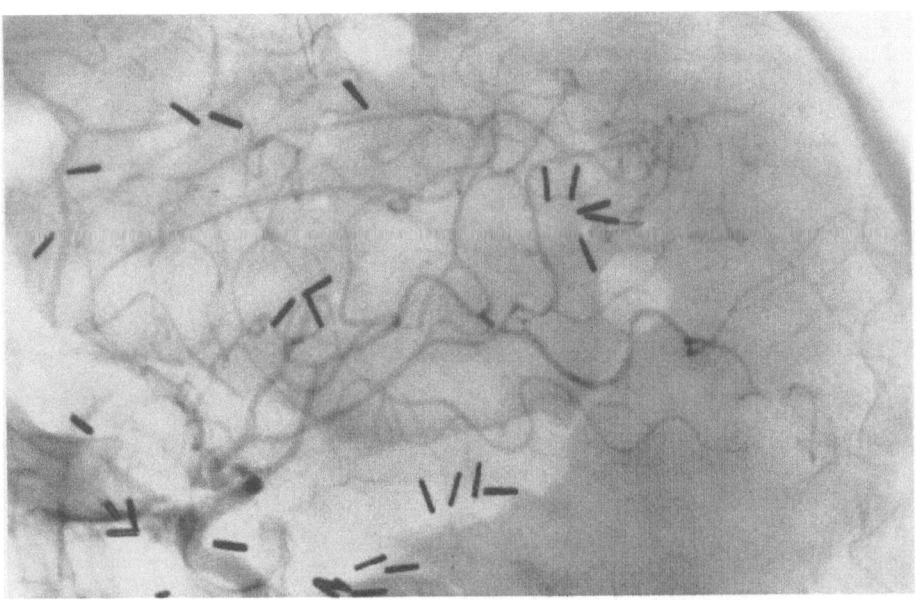

Fig. 105. H. M. S., 34 year old woman, 236/53. *Left internal carotid angiography* (postoperative). The lesion has been removed. Note the improved filling of the anterior cerebral vessels. Condition of the patient was improved at the time of discharge. Minimal weakness of the arm remains

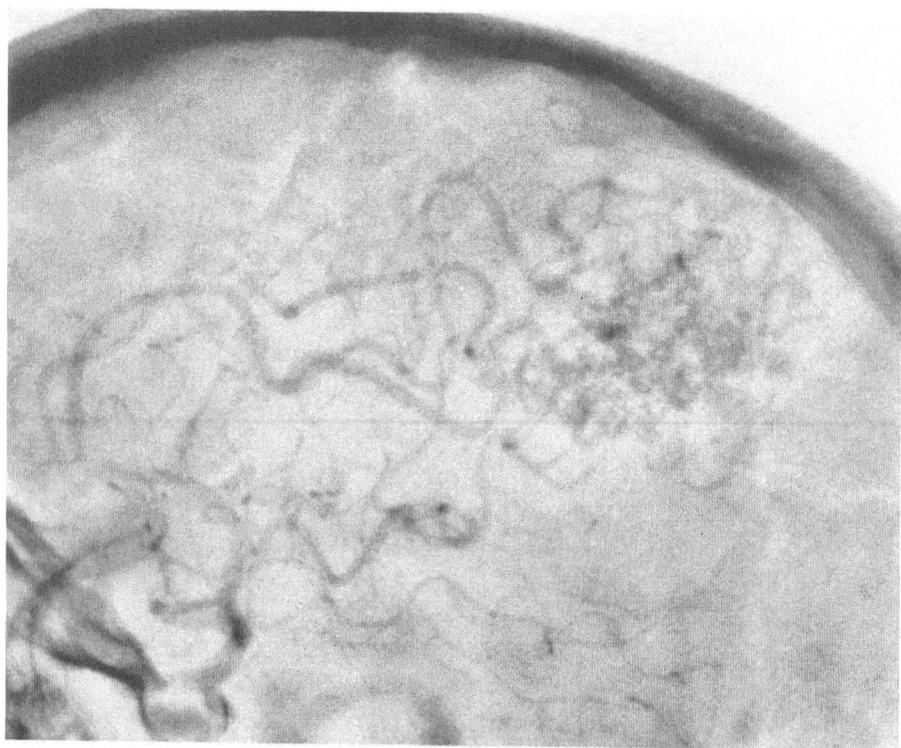

Fig. 106. B. R. J., 29 year old woman, 877/53. Epileptic attacks for five months. *Left internal carotid angiogram* (preoperative). A parietal lesion extending from the medial surface of the hemisphere into the parenchyma. Supply from the pericallosal and callosomarginal arteries. Drainage to the superior longitudinal sinus and to the great vein of Galen

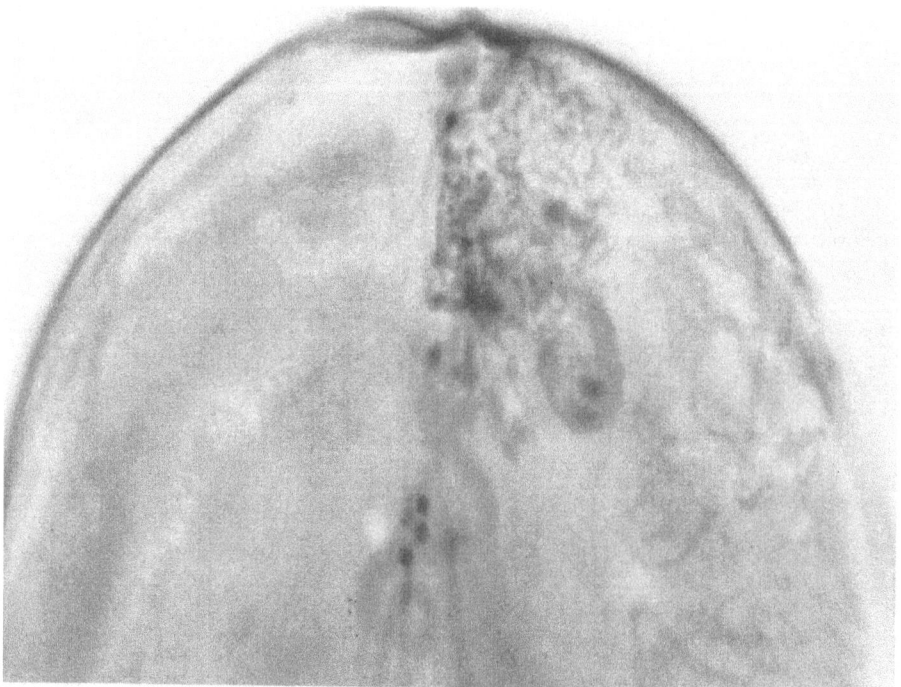

Fig. 107. See text of Fig. 106

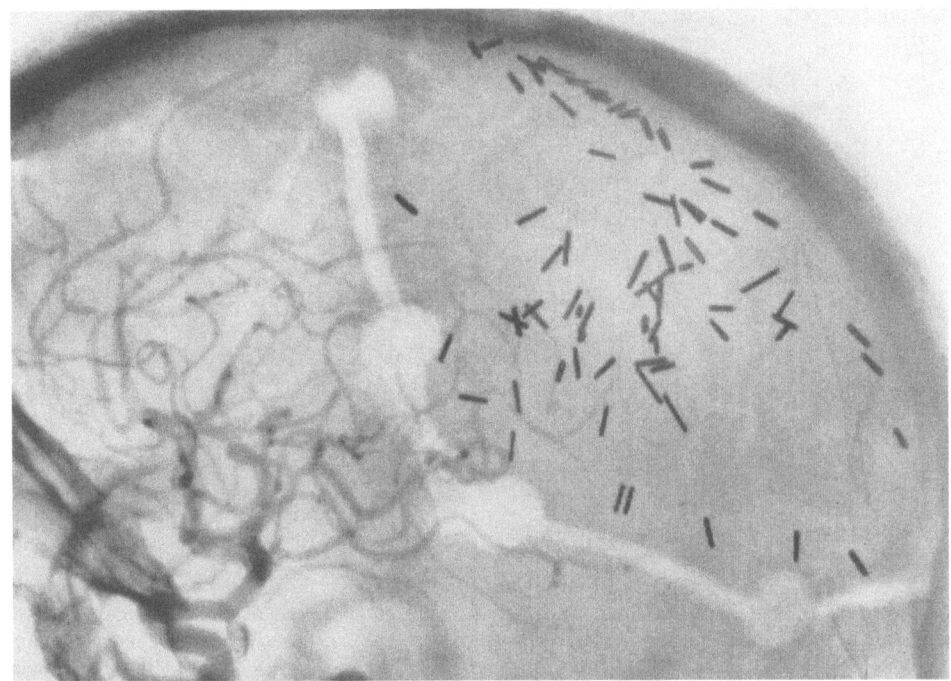

Fig. 108. B. R. J., 29 year old woman, 877/53. *Left internal carotid angiogram* (postoperative). The lesion has been removed. The patient is fully employable. Occasional epileptic attack

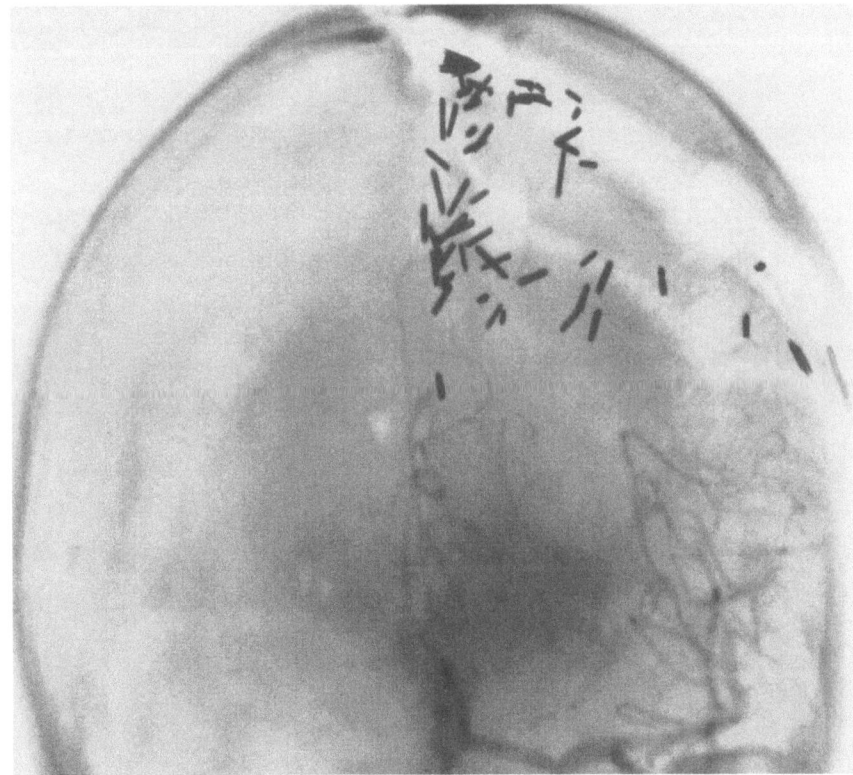

Fig. 109. B. R. J., 29 year old woman, 877/53. *Left internal carotid angiogram* (postoperative). The lesion has been removed. The patient is fully employable. Occasional epileptic attack

78

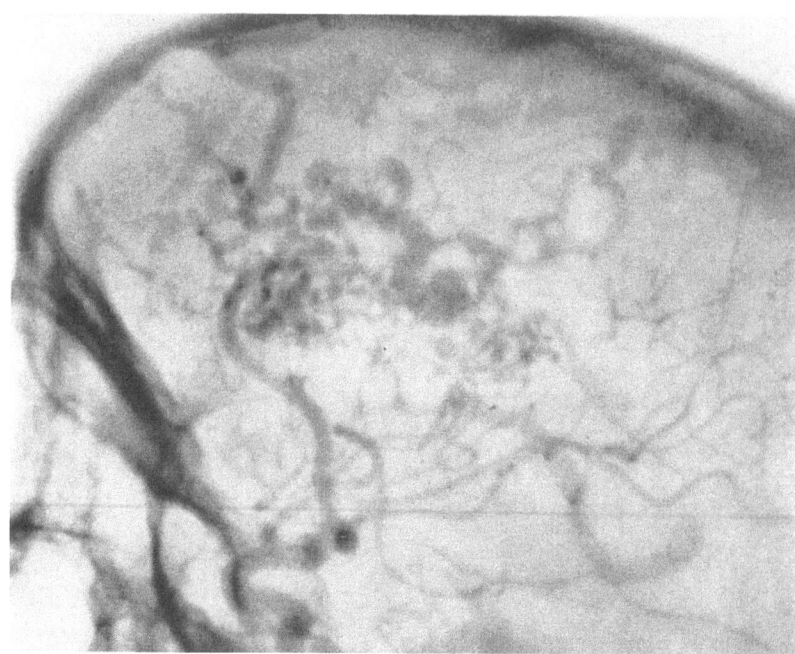

Fig. 110. E. M., 25 year old woman, 1955 (not included in this series). Borderline operability. *Right internal carotid angiogram.* Dilated pericallosal artery which supplies the malformation. Drainage to the superior longitudinal sinus and the internal cerebral vein. Two episodes of intracranial hemorrhage since the age of 17, associated with transient right sided hemiparesis and paresis of the facial musculature. Two epileptic seizures in the preceding year. The operation, undertaken with some reluctance, proved not as formidable as anticipated.

Epileptic attacks have ceased. Excellent clinical result. Postoperative angiograms not yet available

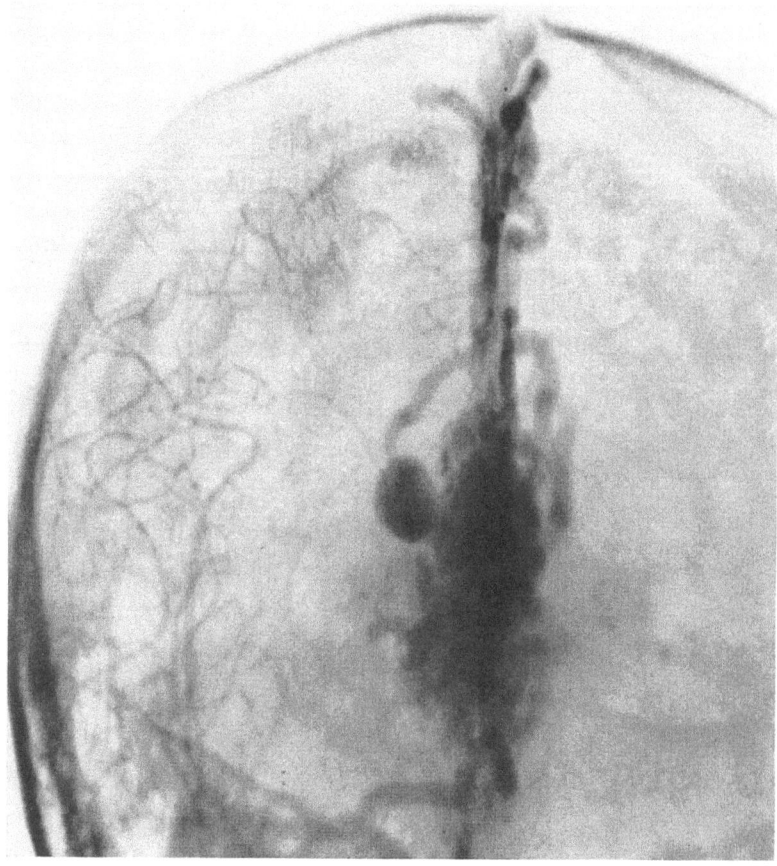

Fig. 111. See text of Fig. 110

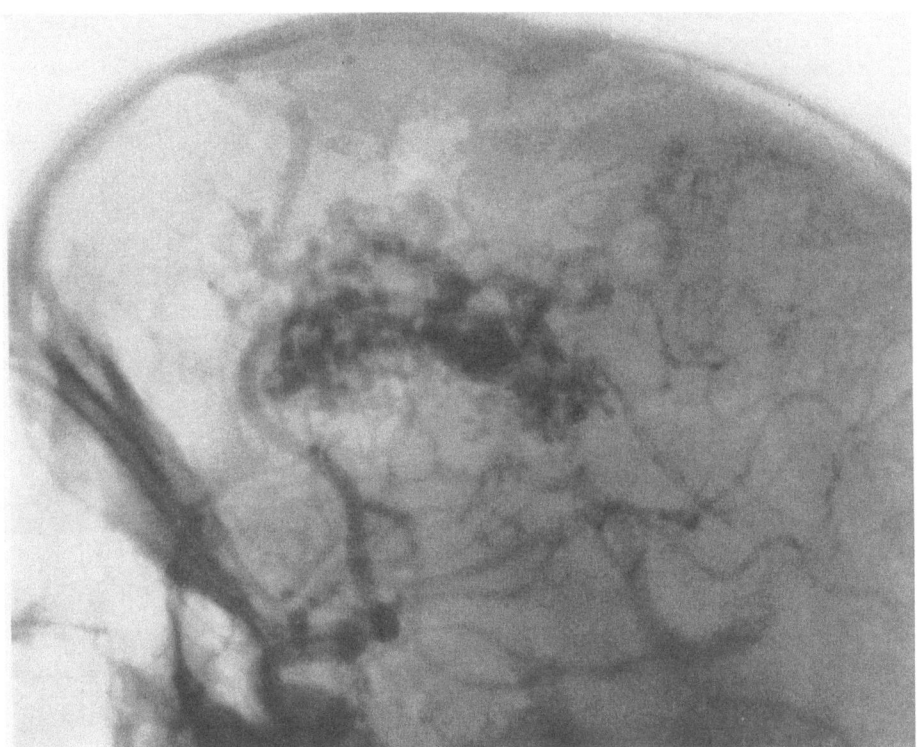

Fig. 112. E. M., 25 year old woman, 1955 (not included in this series). *Left internal carotid angiogram.* The angioma also fills from the left side

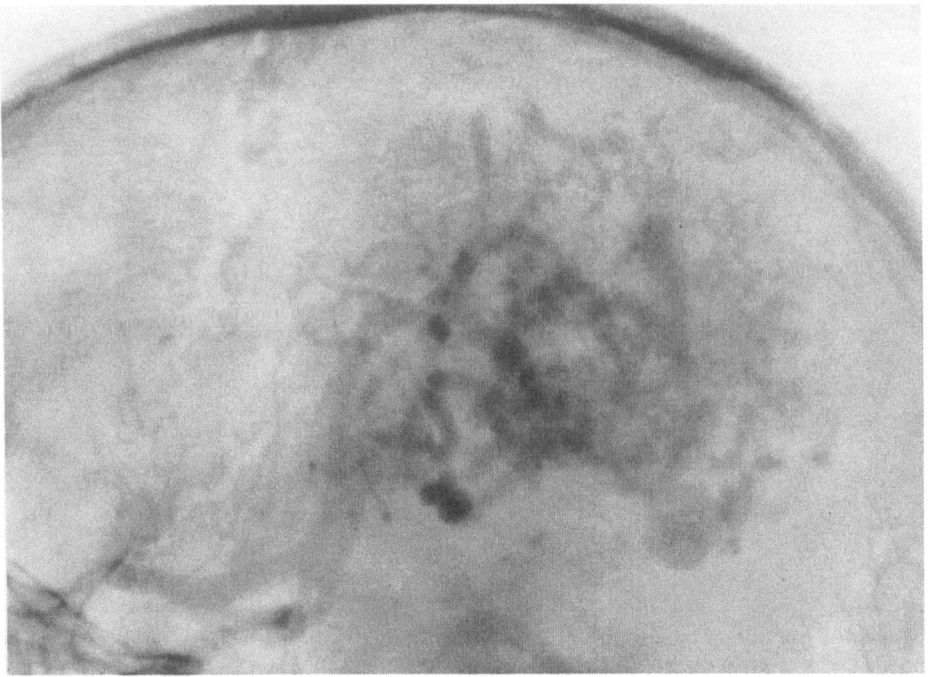

Fig. 113. K. G. V. R. An inoperable lesion. *Right internal carotid angiogram.* The temporal and parietal branches of the middle cerebral artery are dilated and supply the parietal part of the lesion

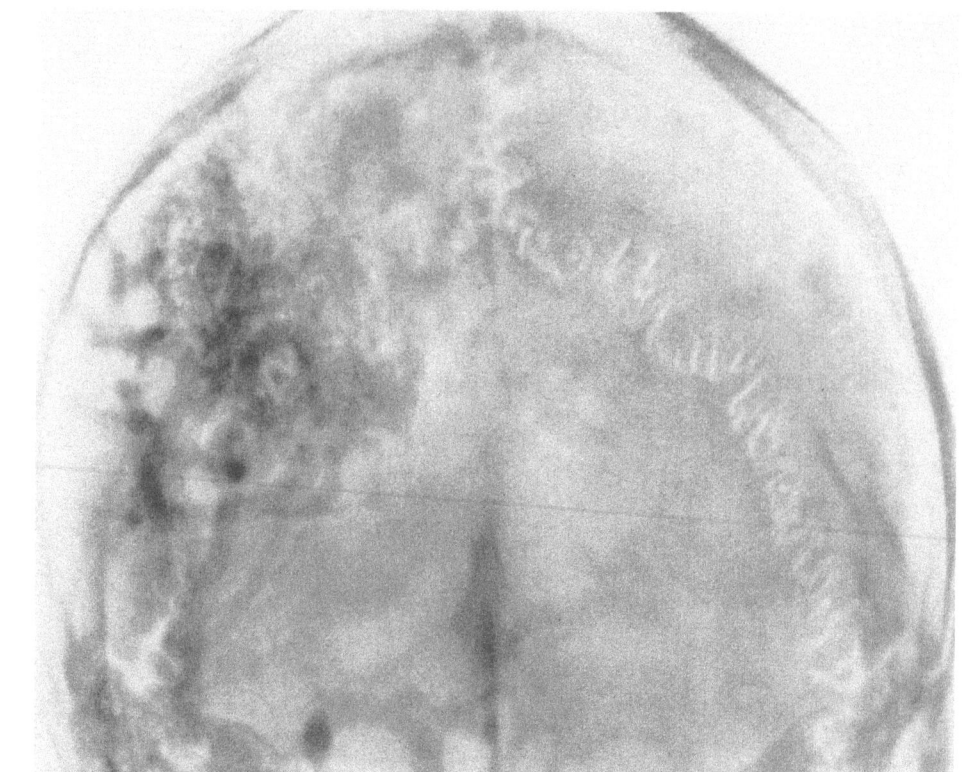

Fig. 114. See text of Fig. 113

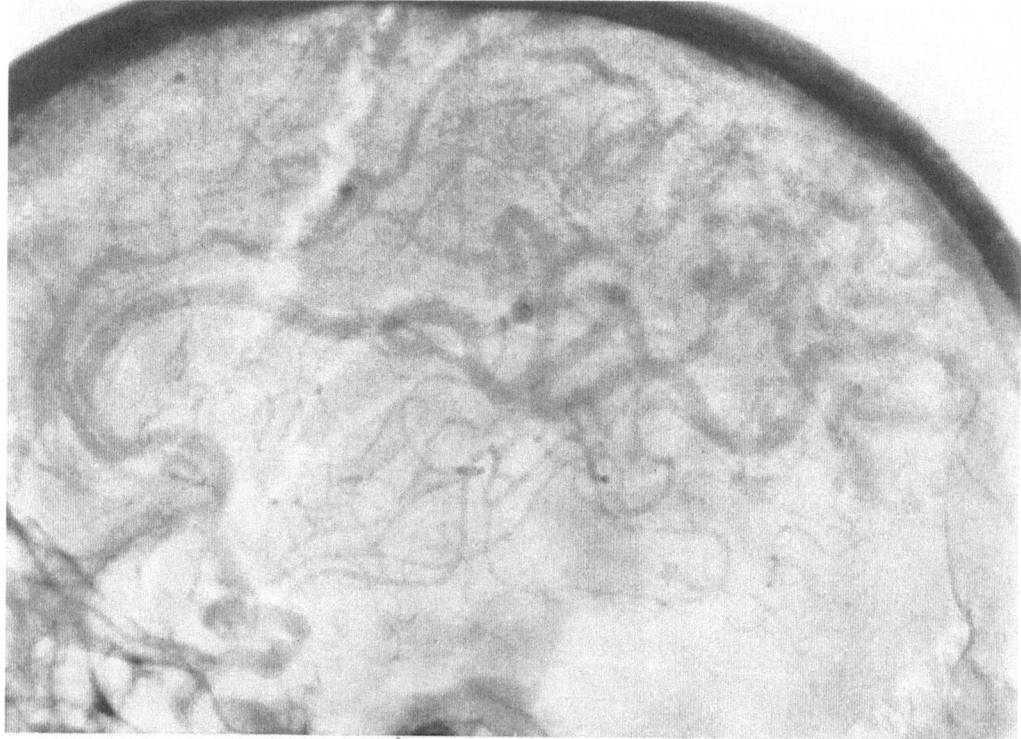

Fig. 115. K. G. V. R. An inoperable lesion. *Left internal carotid angiogram.* Dilatation of the right anterior cerebral artery which receives its supply via the anterior communicating artery. Dilatation of the right peri-callosal and marginal callosal arteries, which supply a cuneiform lesion lying in the midline. Drainage to the superior longitudinal sinus and to the region of the sylvian fissure

many years with no greater inconvenience than an occasional epileptic attack. In the absence of an absolute rule to guide his decision, the surgeon must carefully plot each borderline case against the absissa of danger and the ordinance of benefit (Fig. 110.112).

**Contraindications.** The contraindications to surgery follow the same pattern as with the neoplatic lesions of the brain. Malformations situated in such surgically inaccessible areas as the basal ganglia, capsule, etc. must be avoided. The large lesion, especially that involving both external and middle cerebral arteries in the dominant hemisphere, should be considered a *noli me tangere*. Lesions of the anterior choroidal artery should be viewed with suspicion, and, from a sociological viewpoint, epilepsy of long standing has little to gain from operative intervention (Fig. 113–115). As with malformations situated in the dominant motor area, surgery of the cerebellar lesion must be carefully selected with regard to size and location of the aneurysm. Usually the formidable dimensions of these cerebellar malformations can be readily discerned in the angiogram, but where a reasonable doubt exists as to the extent of the lesion, exploration may be undertaken to vouchsafe the patient every opportunity for surgical benefit. Lesions of the cerebello-pontine angle may at times be explored, but if they are unfavorably situated, the prudent surgeon will refrain from attempting excision.

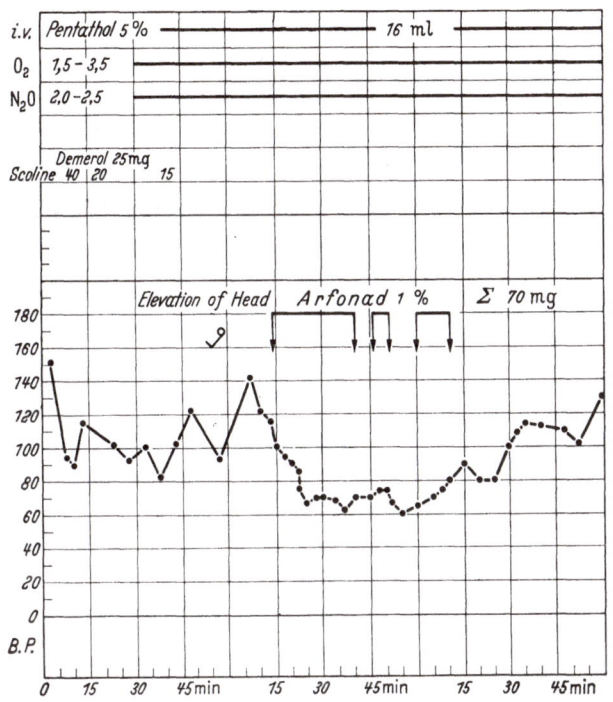

Fig. 116. Anesthetic course under arfonad (hypotension anesthesia). A. E. A., 82/54, a 44 year old woman with a history of atypical right face pain. During the previous six months the patient had also noted that the hair on the right side of her head had turned white and that her right forehead had became depigmented. A right subfrontal lesion was demonstrated by the angiogram

## E. Anesthesia

In the early days of neurosurgery an excision of an arteriovenous aneurysm was a formidable undertaking. The introduction of high spinal anesthesia and hexamethonium was of some benefit to surgery, but both procedures had their shortcomings. Total sympathetic blockage was hazardous and even with proper positioning of the patient, the regulation of blood pressure was difficult. The action of hexamethonium was somewhat unpredictable, with hypersensitivity and resistance often complicating the induction of hypotension.

At present we have come to rely on arfonad (R-2-2222), a ganglionic blocking agent, to effect hypotension. Response of the blood pressure to the intravenous administration of arfonad is rapid and the drug is quickly consumed by the body. A complete cardiological investigation of the patient should be conducted before the use of the drug is contemplated, for, following the administration of arfonad, electrocardiac changes will appear in 5 out of 15 patients, 4 of whom will subsequently show a return to the preoperative pattern by the time of discharge.

Soon after the incision has been made, arfonad is administered by intravenous drip into a leg vessel at the rate of 10–15 drops/minute. The pressure is permitted to fall to 60–70 mm. during the critical phase of the procedure and maintained at hypotensive levels until the time for closure of dura, when the drug is discontinued and the pressure

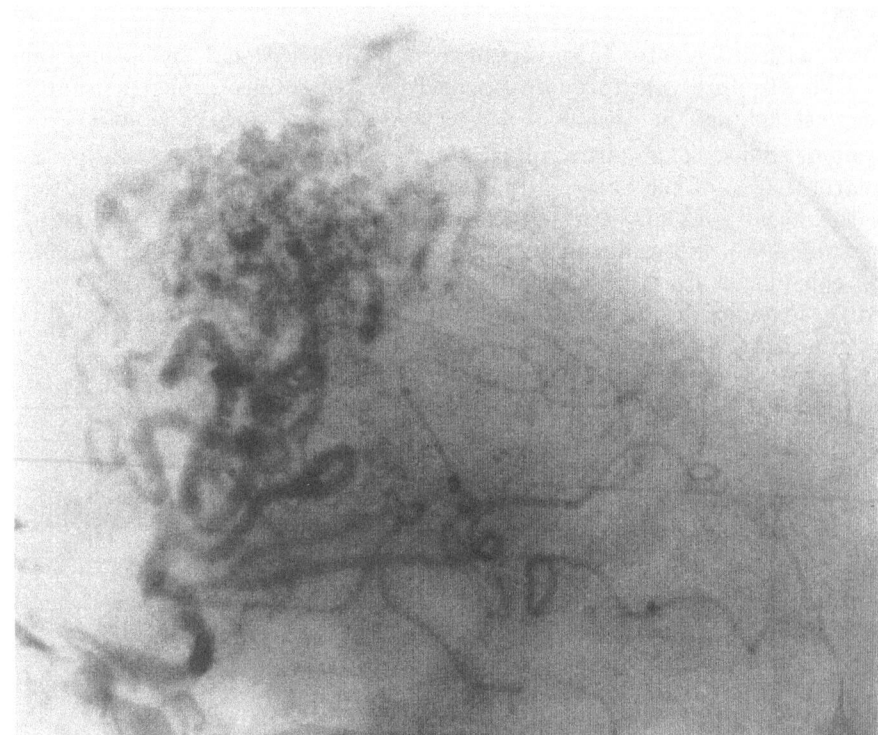

Fig. 117. K. K., 15 year old boy, 340/52. Intracranial bleeding 2 months before surgery. *Left common carotid angiography.* Lesion is supplied by the anterior cerebral and middle cerebral arteries. Drainage to the superior longitudinal sinus and to the base of the skull

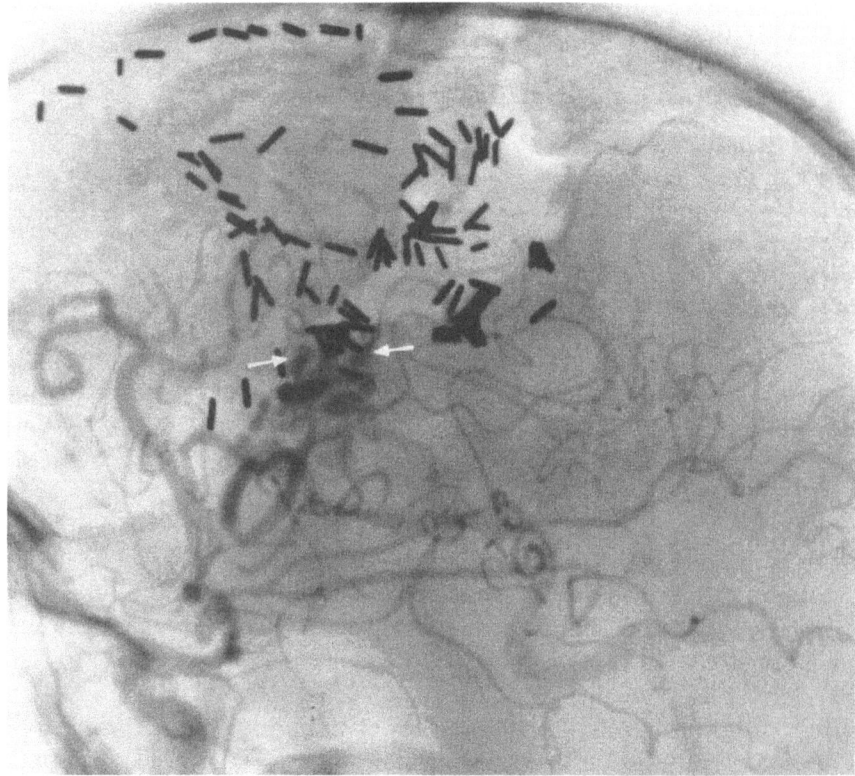

Fig. 118. K. K., 15 year old boy, 340/52. *Left internal carotid angiography.* Some of the malformations remains, fed by the anterior and middle cerebral arteries. Patient is fully employed. No complaints

allowed to rise to the induction level (Fig. 116). It is well to bear in mind that arfonad must not be administered before blood loss is replaced.

Administered preoperatively (100–150 mgm. *per os*) and during surgery (50 mgm. I.M.) to the hypertensive patient, hibernal (chlorpromazine) can diminish the amount of arfonad

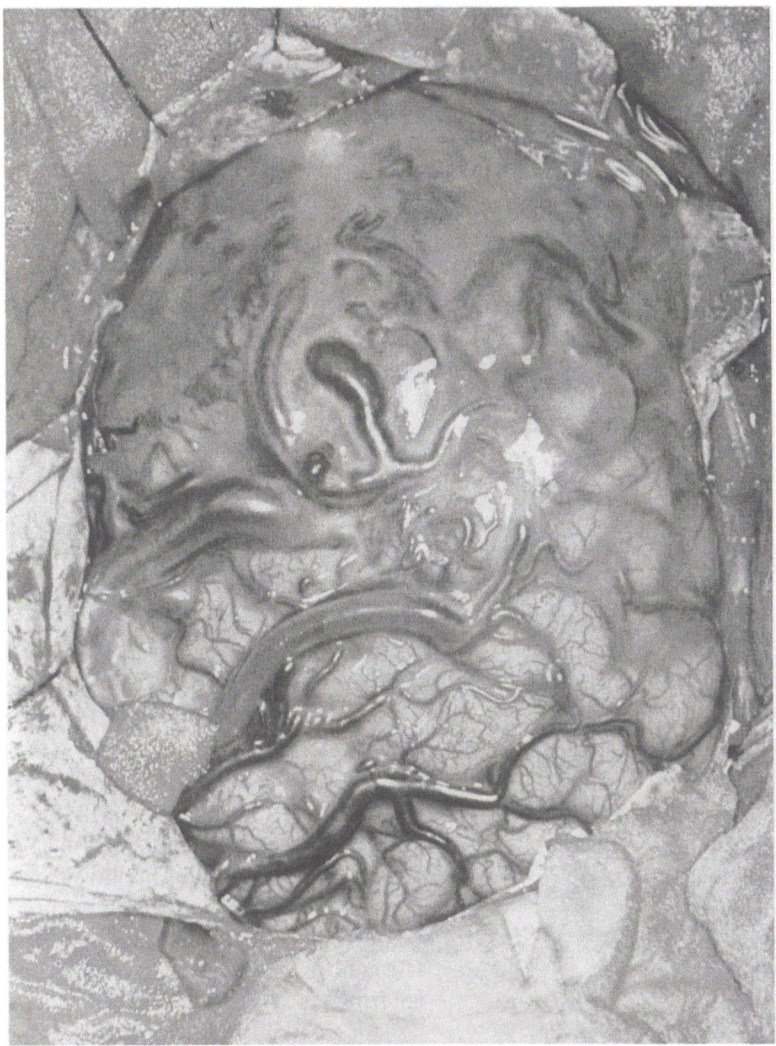

Fig 119. K. K., 15 year old boy, 340/52. Appearance of the arteriovenous aneurysm at surgery

required during the operative procedure, if supplemented by an elevation of the head of the patient.

Details of the experience of this clinic with the hypotensive drugs may be found in a previous study[17].

# F. Operative Technique

## 1. Cerebral Lesions

The exposure must be planned so that all of the feeding arteries visualized in the angiogram will be within easy access in the operative field (Fig. 117, 118). When branches of the anterior cerebral artery contribute, the margin of the longitudinal sinus should be exposed to allow for ligation of these branches as they course in the medial surface

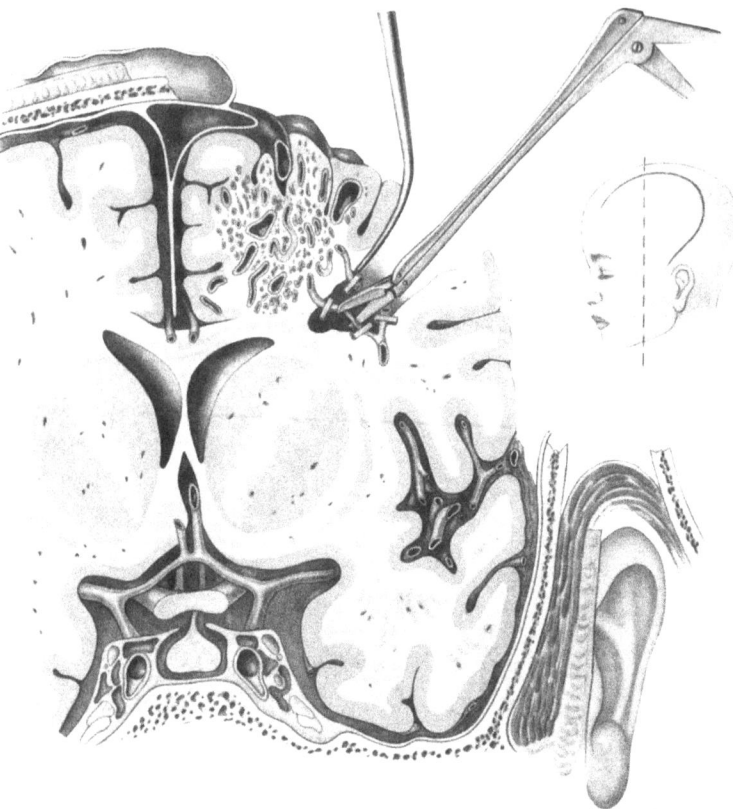

Fig. 120. K. K., 15 year old boy, 340/52. Removal of aneurysm. The dissection is gradually carried deeper and the blood vessels are occluded by silver clips as they enter the lesion

of the hemisphere. The frequent communications between vessels of the dura and the aneurysm require that the reflection of the dura be made with caution to avoid traumatizing these vessels before they can be clipped and coagulated.

After the lesion is exposed (Fig. 119), the pia-arachnoid should be incised at the periphery of the malformation. All arteries entering the area are clipped and divided, together with such small veins as are encountered. The larger veins, which are the main channels of drainage, should be left intact until a later stage of the dissection.

By blunt dissection aided by the cupped spatula, the vessels penetrating the aneurysm are exposed (Fig. 120). The cuneiform lesion extends in the direction of the ventricle with the main blood supply entering beneath the cortex in the form of small, thin-walled arteries. These fragile vessels should be clipped rather than cauterized, for if traumatized they will retract into the white matter, where they are troublesome to find.

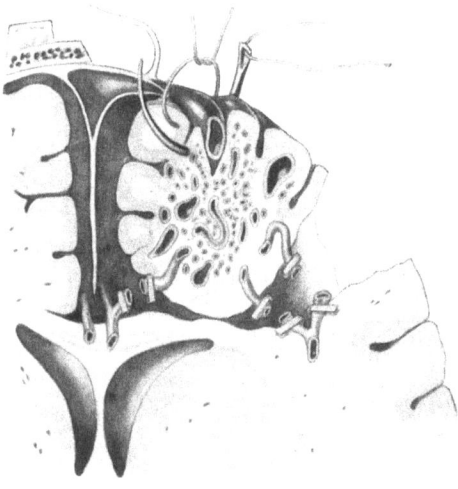

Fig. 121. K. K., 15 year old boy, 340/52. After the arterial supply has been occluded, the large veins are ligated and the malformation removed

As the dissection proceeds, several of the larger veins should be divided to facilitate the mobilization of the lesion, but at least one vein must be left undisturbed to continue drainage. When the malformation has been sufficiently mobilized, it is grasped in a special broad-faced (Brüning) clamp which compresses the major part of the lesion, and by gentle traction the deeper blood vessels can be exposed and clipped. At times it may be easier to ligate the vascular tissue of the apex with a suture. The remaining vein can then be divided and the lesion removed (Fig. 121).

If the procedure has gone well, the operative field will be virtually bloodless at this point (Fig. 122). The blood pressure is then permitted to rise, and after further assurance of complete hemostasis at normal pressure levels, the dura should be closed without drainage.

## 2. Cerebellar Lesions

Hypotensive anesthesia is an especial prerequisite for surgery of the posterior fossa lesion.

A suboccipital exposure should be performed bilaterally. The foramen magnum may be enlarged and at times an occipital flap may be required if the lesion is situated in the region of the incisura. The dura is opened with practised circumspection, lest the large veins be traumatized and the surgeon committed to a procedure more extensive

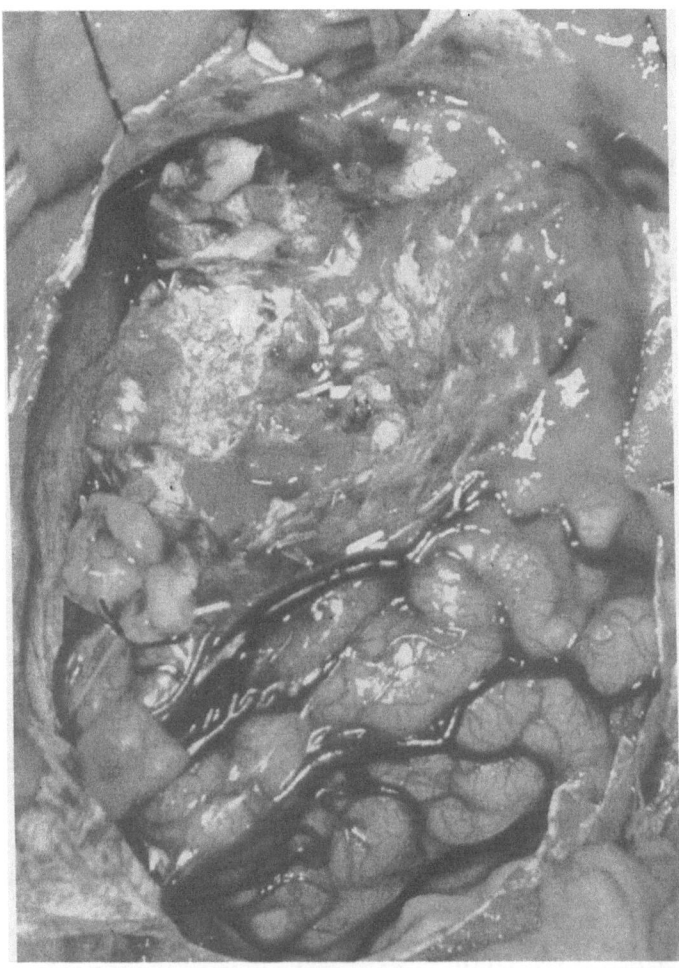

Fig. 122. K. K., 15 year old boy, 340/52. The lesion has been removed. Note the appearance of the veins in contrast to Fig. 119

than he would otherwise have contemplated. At this point the decision concerning the advisability of extirpation must be made, guided by the location and dimensions of the lesion.

If extirpation is elected, it is usually best to begin mobilization at the mid-line, since the large draining veins which enter the transverse sinus are very friable. The afferent artery, identified by its increased diameter, should be clipped and divided, and, proceeding from the surface of the lesion at its periphery, the dissection into the depths of the lobe is commenced. As with the malformations of the cerebrum, all vessels should be clipped rather than coagulated. Furthermore, at least one draining vein leading into the transverse sinus must be retained to insure drainage. Mobilization of the deep vessels may often be facilitated by grasping the tentorium with a hemostat and raising it upward. By this manoevre, bleeding from the deep vessels can be controlled and the vessels brought within the grasp of the operator.

# References

[1] BAILEY, P.: Intracranial Tumors, 2nd edit. Springfield, Illinois: Ch. C. Thomas 1948.

[2] BARUK, H.: Migraines d'apparence psychogenique suivies d'epilepsie jacksonienne dans un cas d'angiome cérébral. Encéphale **26**, 42 (1931).

[3] BERGSTRAND, H., H. OLIVECRONA u. W. TÖNNIS: Gefäßmißbildungen und Gefäßgeschwülste des Gehirns. Leipzig: Georg Thieme 1936.

[4] BODREY, E., and E. R. MILLER: Arteriovenous fistula of great cerebral vein and the circle of WILLIS. Report of 2 patients treated by ligation. Arch. of Neur. **62**, 778 (1949).

[5] CALLANDER, C. L.: Study of arteriovenous fistula with an analysis of 447 cases. Ann. Surg. **71**, 428 (1920).

[6] COENEN, H.: Die Geschwülste. Ein Beitrag in M. KIRSCHNER u. O. NORDMANN, Die Chirurgie, Bd. 2, S. 74. Berlin: Urban & Schwarzenberg 1930.

[7] CUSHING, H., and P. BAILEY: Tumors arising from the bloodvessels of the brain. Springfield, Illinois: Ch. C. Thomas 1928.

[8] DANDY, W. E.: Arteriovenous aneurysms of the brain. Arch. Surg. **17**, 190 (1928).

[9] DANDY, W. E.: Surgery of the Brain. In: LEWIS' Practice of Surgery, vol. XII. Hagerstown, Maryland: W. F. Prior 1945.

[10] DANDY, W. E., and R. H. FOLLIS: On pathology of carotid cavernous aneurysm (pulsating exophtalmos). Amer. J. Ophthalm. **24**, 365 (1941).

[11] DELENS, E.: De la Communication de la Carotide interne et du sinus Caverneux. Thèse de Paris 1870 (quoted in DANDY and FOLLIS).

[12] DEUTSCH, L., and J. FRIEDMANN: Über ophthalmische Migräne bei Gefäßmißbildung (arteriovenöse Aneurysmen). Dtsch. Z. Nervenheilk. **146**, 199 (1938).

[13] DOLIETIS, K.: Radical removal of arteriovenous aneurysms in the masseter muscle. Acta chir. scand. (Stockh.) **98**, 177 (1949).

[14] ELKIN, D. C.: Circoid aneurysm of the scalp. Ann. Surg. **80**, 332 (1924).

[15] FULTON, J. F.: Textbook of Physiology, 15th edit., p. 848–852. Philadelphia, Pennsylvania: W. B. Saunders Company 1947.

[16] GERBODE, F., E. HOLMAN, E. DICKENSON and R. SPENCER: Arteriovenous fistulas and arterial aneurysms. Surgery **32**, 259 (1952).

[17] GORDON, E., and J. LADENHEIM: Controlled hypotension in neurosurgery. Acta chir. scand. (Stockh.) **109**, 488 (1955).

[18] HAMBY, W. B.: Intracranial aneurysms, p. 428. Springfield, Illinois: Ch. C. Thomas 1952.

[19] HOFFMAN, J.: Krankenvorstellung. Münch. med. Wschr. 1898, 1159.

[20] HOLMAN, E.: The physiology of an arteriovenous fistula. Arch. Surg. **7**, 64 (1923).

[21] HOLMAN, E.: Roentgenologic kymographic studies of the heart in the presence of an arteriovenous fistula and their interpretation. Ann. Surg. **124**, 920 (1946).

[22] HÖÖK, O., L. WERKÖ and G. ÖHRBERG: Personal communication.

[23] HUSBY, J., G. NORLÉN and J. PETERSÉN: Electroencephalographic findings in intracranial, arterial and arteriovenous aneurysms and subarachnoid hemorrhages. Acta psychiatr. (Københ.) **28**, 387 (1953).

[24] HYLAND, H. H., and R. P. DOUGLAS: Cerebral angioma arteriale. Arch. of Neur. **40**, 1220 (1938).

[25] ISENSCHMID, R.: Die klinischen Symptome des cerebralen Rankenangiom. Münch. med. Wschr. 1912, 243.

[26] JAEGER, J. R., R. P. FORBES and W. E. DANDY: Bilateral congenital arteriovenous communication. Tr. Neurol. a. Neurol. A **63**, 173 (1937).

[27] KETY, S S.: Circulation and metabolism of human brain in health and disease. Amer. J. Med. **8**, 205 (1950).

[28] KÖRTE, W.: Beitrag zur Lehre vom Angioma arteriale racemosum. Dtsch. Z. Chir. **13**, 24 (1880).

[29] LANGE-COSACK, H.: Handbuch der Chirurgie von KIRSCHNER u. NORDMAN. 1948. Quoted in [63].

[30] LEVINE, S. A.: Clinical Heart Disease, 2nd edit. Philadelphia, Pennsylvania: W. B. Saunders Company 1942.

[31] LIMA, P. A.: Cerebral Angiography. London: Oxford Univ. Press 1950.

[32] LINDGREN, E.: In Handbuch der Neurochirurgie, herausgeg. von H. OLIVECRONA u. W. TÖNNIS, Bd. II, Röntgenologie. Berlin: Springer 1954.

[33] MACKENZIE, I.: The clinical presentation of the cerebral angioma. A Review of 50 cases. Brain **76**, 184 (1953).

[34] MELENEY, F.: A pathological study of a case of circoid aneurysm. Surg. etc. **36**, 547 (1923).

[35] MURPHY, J. P.: Cardiovascular Disease. Chicago, Illinois: Year Book Publ. 1954.

[36] NORLÉN, G.: Arteriovenous aneurysms of the brain. J. of Neurosurg. **6**, 475 (1949).

[37] OLIVECRONA, H.: Arteriovenous aneurysm i hjärnan. Nord. Med. **41**, 843 (1949).

[38] OLIVECRONA, H.: Gli aneurismi arterio-venosi del cervello. Minerva med. (Torino) Suppl. **1950**, 118.

[39] OLIVECRONA, H.: Die arteriovenösen Aneurysmen des Gehirns. Dtsch. med. Wschr. **1950**, 1169.

[40] OLIVECRONA, H., and J. RIIVES: Arteriovenous aneurysms of the brain. Arch. of Neur. **59**, 567 (1948).

⁴¹ OSLER, W.: Remarks on arteriovenous aneurysm. Lancet **1915**, 949.

⁴² PADGETT, D. H.: The development of the cranial arteries in the human embryo. Washington, D.C.: Carnegie Institute in Contributions to Embryology **32**, 207 (1948).

⁴³ POTTER, J. M.: Angiomatous malformations of the brain: their nature and prognosis. Ann. Roy. Coll. Surg. **16**, 227 (1955).

⁴⁴ RAY, B. S.: Cerebral arterio-venous aneurysms. Surg. etc. **73**, 615 (1941).

⁴⁵ REID, M. R.: Studies of abnormal arteriovenous communications, acquired and congenital. I. A report of a series of cases. Arch. Surg. **10**, 601 (1925).

⁴⁶ REID, M. R.: Abnormal arteriovenous communications, acquired and congenital. II. The origin and nature of arteriovenous aneurysms, circoid aneurysm and simple angiomas. Arch. Surg. **10**, 996 (1925).

⁴⁷ REID, M. R.: Abnormal arterial communications, acquired and congenital. III. The effects of abnormal arteriovenous communications of the heart, blood vessels and other structures. Arch. Surg. **11**, 25 (1925).

⁴⁸ REID, M. R., and J. McGUIRE: Arteriovenous aneurysms. Ann. Surg. **198**, 649 (1938).

⁴⁹ REINHOFF, W. F.: Congenital arteriovenous fistula, an embryologic study with report of case. Bull. Johns Hopkins Hosp. **35**, 271 (1924).

⁵⁰ RUGGIERO, G., and F. CASTELLANO: Carotid-cavernous aneurysm. Acta radiol. (Stockh.) **37**, 121 (1952).

⁵¹ RÖTTGEN, P.: Weitere Erfahrungen an kongenitalen arterio-venösen Aneurysmen des Schädel-innern. Zbl. Neurochir. **2**, 18 (1937).

⁵² SCHEINBERG, P., and H. JAYNE: Factors influencing cerebral blood flow and metabolism. Circulation (New York) **5**, 225 (1952).

⁵³ SCHEINKER, I. M.: Neurosurgical Pathology. Springfield, Illinois: Ch. C. Thomas 1948.

⁵⁴ SCHLESINGER, B.: The venous drainage of the brain. Brain **62**, 274 (1939).

⁵⁵ SCHMIDT, C. F.: The Cerebral Circulation in Health and Disease. Springfield, Illinois: Ch. C. Thomas 1950.

⁵⁶ SHENKIN, H. A., E. B. SPITZ and S. S. KETY: Physiological studies of arteriovenous anomalies of the brain. J. of Neurosurg. **5**, 165 (1948).

⁵⁷ SONNTAG, E.: Das Rankenangiom sowie die genuine diffuse Phlebarteriektasie und Phlebektasie. Erg. Chir. **11**, 99 (1919).

⁵⁸ SORGO, W.: Weitere Mitteilungen über Klinik und Histologie des kongenitalen arterio-venösen Aneurysmas des Gehirns. Zbl. Neurochir. **3**, 64 (1938).

⁵⁹ STEINHEIL, S. O.: Über einen Fall von Varix aneurysmaticus im Bereich der Gehirngefäße. Inaug.-Diss. Würzburg: F. Fromme 1895.

⁶⁰ STRABOS, R. R. J., and L. MOUNT: Problems relating to treatment of intracranial aneurysms by carotid ligation. Arch. of Neur. **69**, 118 (1953).

⁶¹ STREETER, G. L.: The developmental alterations in the vascular system of the brain of the human embryo. Washington, D.C.: Carnegie Institute, in Carnegie Publ. **1918**, No 271, 9.

⁶² SUGAR, O.: Pathologic anatomy and angiography of intracranial vascular anomalies. J. of Neurosurg. **8**, 3 (1951).

⁶³ TÖNNIS, W., u. H. LANGE-COSACK: Klinik, operative Behandlung und Prognose der arterio-venösen Angiome des Gehirns und seiner Häute. Dtsch. Z. Nervenheilk. **170**, 460 (1953).

⁶⁴ TÖNNIS, u. W., W. SCHIEFER: Zur Frage des Wachstums arterio-venöser Angiome. Zbl. Neurochir. **3**, 145 (1955).

⁶⁵ BOGAERT, L. VAN: Pathology of angiomatosis. Acta neurol. et psychiatr. belg. **50**, 525 (1954). Quoted in MURPHY.

⁶⁶ VIRCHOW, R.: Die krankhaften Geschwülste, Dd. 3, S. 306-496. Berlin 1863.

⁶⁷ WAGNER, A.: Über das arterielle Ranken-Angiom an den oberen Extremitäten. Beitr. klin. Chir. **11**, 49 (1894).

⁶⁸ WARD, E. E., and H. BAYARD: Congenital arteriovenous fistulas in children. J. of Pediatr. **16**, 746 (1940).

⁶⁹ WARREN, J. V.: Blood volume in patients with arteriovenous fistulas. J. Clin. Invest. **30**, 220 (1951).

⁷⁰ WARREN, J. V., J. L. NICKERSON and D. C. ELKIN: Cardiac output in patients with arteriovenous fistulas. J. Clin. Invest. **30**, 210 (1951).

⁷¹ WECHSLER, I. S., S. W. GROSS and I. COHEN: Arteriography and carotid artery ligation in intracranial aneurysms and vascular malformation. J. Neurol., Neurosurg. a. Psychiatry **14**, 25 (1951).

⁷² WYBURN-MASON, R.: Arteriovenous aneurysm of midbrain and retina, facial naevi and mental changes. Brain **66**, 163 (1943).

⁷³ Circulation of the Brain and Spinal Cord. Vol. 18 of the Research Publications, Association for Research in Nervous and Mental Diseases. Baltimore, Maryland: Williams Wilkes 1938.

⁷⁴ Röntgen Department, Serafimerlasarettet: Personal communication.

# Author Index

The page numbers given in *italics* refer to the literature.

# Subject Index